Beyond the Spring

Cordelia J. Stanwood of Birdsacre

"One touch of nature makes
the whole world kin."

Chandler S. Richmond

Nine Golden- crowned Kinglets (CJS photo).

Beyond the Spring

Cordelia J. Stanwood of Birdsacre

by

Chandler S. W. Richmond

New and Revised Edition

with photographs from the Stanwood Collection

THE LATONA PRESS
1989

DOWNEAST GRAPHICS & PRINTING, INC.
2008

10 9 8 7 6 5 4 3

Library of Congress Catalogue Card Number: 88-083667

ISBN — Hard cover: 0-932448-02-X
ISBN — Soft cover: 0-932448-03-8

Book designed by George Orzel

Printed in the United States of America

DOWNEAST GRAPHICS & PRINTING, INC.

THE LATONA PRESS

1889

To

IDELLA STANWOOD LAWTON

whose willingness to share
her memories brought warmth
and vitality to the writing
of her sister's life.

1986

Contents

Illustrations

(the photographers, when known, are identified in the captions)

Preface

A song sparrow caught by a cat in 1947 put me on the path that led eventually to Birdsacre and to this book. I brought it home in a shoebox, with no experience with injured birds and no expectation that it would survive. But survive it did. I built an indoor-outdoor home for it; in a month it began to sing, and it lived, though disabled, for seven years more. Since then I have been increasingly involved in the care and rehabilitation of injured and orphaned birds, and it has been my rare privilege to work intimately with 121 species, from the ruby-throated hummingbird to the bald eagle.

Because of my deepening interest in birds I became a charter member of the Cordelia Stanwood Bird Club in Ellsworth, Maine, in 1955; and because of my membership in that club I was asked to help go through Miss Stanwood's papers when she donated them to the organization. I vividly remember those short cold winter days of 1958, when I recognized at once from her voluminous research records that here was a woman worthy of the most serious consideration. She epitomized my own developing philosophy, not only for what she achieved as a naturalist but for who she was as a person.

This book is the result of twenty years of commitment to discovering who Cordelia Stanwood was.

I have drawn on four major manuscript sources in my study of the Lady of Birdsacre. Most significant for me were Cordelia Stanwood's field notebooks, spanning nearly fifty years, a treasure house of detailed information and sensitive interpretation. A second resource was the reams of manuscripts, published and unpublished, that constitute Cordie's literary and scientific output. They provided me with an almost overwhelming choice of material to illustrate the quality and style of her writing. Third is the scrapbook of correspondence with editors, authors, photographers, and naturalists, furnishing strong evidence of her authority among ornithologists, as well as particulars about her career as a free-lance writer and photographer. Finally, I found much of what I needed about her personal life in her autobiographical and genealogical papers.

Another kind of resource has been the collection of over nine hundred of Cordelia Stanwood's original 5x7 photographic plates. Miss Stanwood had donated more than 500 of these to Acadia National Park, and they had been presumed lost in the Bar Harbor fire of 1947. William Townsend discovered this treasure in a Bar Harbor basement in 1974, and the National Park Service has returned them to the Museum at Birdsacre. These plates have recently been catalogued and indexed, with a generous sampling of negative reproductions and positive prints. This project was assisted by a grant from the Maine State Commission for the Arts and Humanities and carried out through the generosity and talents of Frank and Dorothea Stoke and Ada Graham. In addition to Cordelia's own negative plates and prints I have drawn on the homestead collection of Stanwood family portraits.

All these manuscripts and photographs are copyrighted by the Cordelia Stanwood Wildlife Foundation and are in the permanent collection at the Homestead Museum.

I would like to thank the many persons who have assisted me in the preparation and writing of this book. I am particularly grateful to Lucia Smith Merritt for undertaking the massive task of sorting, cataloging, and filing the manuscript and printed material that Cordelia Stanwood accumulated throughout her lifetime and donated to the Bird Club.

I am indebted to many others whose advice, encouragement, critical judgement, and active assistance have been invaluable to me. Conversations with those who still had vivid memories of Miss Stanwood at various stages of her life helped me to round out the character and personality of this gifted and complex individual. Chief among these was Idella Stanwood Lawton, whose infallible memory linked time and place in her sister's story whenever the records were unclear or inadequate. Others who contributed generously were Effie Anthony, Berenice DeLaite, June Forsythe, Vera Holt, Agnes Huston and Hazel Walker, all friends of Cordelia Stanwood. I am also indebted to the late Dr. Clarence C. Little and Reverend Margaret Henrichsen and to Dr. Thomas Williams, all of whom read an early version of this manuscript and offered constructive criticism which led to a much broader and more definitive biography than I had originally planned.

No one could be more fortunate than I to have had the guidance and professional support of Marion and David Stocking, in preparing this version of the book, and also of Ada and Frank Graham Jr.

Last, but by no means least, was the help in preparing the typescript that Gladys Lord, Phyllis Canfield, and my wife Marion so diligently performed.

"There is nothing either good or bad but thinking makes it so." An unshakable kind of positive thinking by my wife Marion during the twenty years of struggle through the ups and downs of authorship has had much to do with dispelling the bad and emphasizing the good reflected in this biography.

From the beginning I was motivated by the earnest desire to create a sympathetic and unbiased study of Cordelia Stanwood that would not only pay tribute to her accomplishments but dissolve the misunderstandings about this uniquely gifted woman. What I have written is but the frame for the portrait that emerges through her own words and thoughts and the experiences that shaped her life.

Chandler S. Richmond
Ellsworth, Maine
22 August 1978

Chronological Chart The Life of Cordelia Stanwood

1865	Born, August 1, at the Stanwood homestead on Beckwith Hill, Ellsworth, Maine.
1879	Went to Providence, R.I. to live with Aunt Cordelia Johnson and attend public school.
1882	Graduated from Federal Street Grammar school. Joined First Baptist Church, December 31.
1886	Graduated from Providence Girls High School.
1887	Attended Teacher Training School.
1887-1893	Taught third to fifth grades in the Providence school system. Principal of Plain Street School, 1892-1893.
1890	Spent the summer on Cape Breton Island with her maternal grandparents at Eskasoni. Visited Louisburg.

1891	Attended summer school at Martha's Vineyard.
1892	Attended summer school at Glens Falls, N.Y. Death of her sister Maude.
1893-1894	Attended Normal Arts School in Boston, completing the two-year course in one year.
1894-1896	Assistant Supervisor of Drawing in the Springfield, Mass. school system.
1896-1898	Supervisor of Drawing in the Greenfield, Mass. school system.
1898-1899	Sixth-grade teacher in William H. Lincoln School, Brookline, Mass.
1899-1901	Teacher in the Quincy School, Poughkeepsie, N.Y., a training school for Vassar College.
1901	Attended summer school at Harvard University.
1901-1903	Supervisor of Drawing in Plymouth and Braintree, Mass.
1903-1904	Supervisor of Drawing in Braintree and Marion, Mass.
1904	Nervous breakdown and return to homestead in Ellsworth.
1905	Began intensive study of natural history, particularly birds.
1909-1940	Extensive writing and publication.
1914	Death of her father.
1916-1926	Became professional photographer.
1932	Death of her mother.
1950	Death of her sister Maria.
1958	Cordelia Stanwood died in Ellsworth, Nov. 20.

The Hermit Thrush

When the dew is falling,
  When the daylight wanes,
Then my soul respondeth
  To thy haunting strains,
O hermit thrush!

Did the heavens open
  Shining portals wide,
And a joyous singer
  Stray here to abide?
Dear hermit thrush!

Wildest voice of woodland,
  Sweetest song of praise;
Listening to thy hymning
  Let me spend my days!
Blythe hermit thrush!

Cordelia J. Stanwood
June 8, 1909

Prologue

It was a clear, crisp December day, and I was in the Winter House at Birdsacre preparing food for my invalid owls and broad-winged hawks while outside the ducks and geese clamored for corn. Time and change were on my mind. Only the day before I had been obliged to acknowledge my seventieth birthday. Through the windows in the north wall I looked down upon the ice-covered surface of McGinley Pond where two girls were skating. For me at seventy the cold outside was a vascular hazard, but to these children it was nothing.

Some quick mental arithmetic made it easy to slip back one hundred years to a cold December day in 1876 and visualize eleven-year-old Cordie Stanwood skating with her sister Maria on the frog pond above the house even as these two girls were skating here today on McGinley Pond. The comparison brought home to me how much I had been juggling time and events these past twenty years while traveling back and forth from one century to another to record the lifetime of a woman who had so captivated my own attention and admiration that I felt impelled to tell her story so others might come to know her as I do.

Also, I felt a compulsion to refute the misconceptions about Cordelia Stanwood that had long prevailed among those who saw her only as a strange, eccentric personality who lived alone and remained aloof from the usual patterns of social behavior. I had no desire to make her appear like others by minimizing or rationalizing her oddities, for Cordie Stanwood was wonderfully different and this very difference had made her the sensitive interpreter of nature we come to know through her writing.

Many have asked how I became so involved in the life story of Cordelia Stanwood, and I, when most exasperated by the slowness with which that story has unfolded, have asked myself the same question time and again. No single answer would suffice, no simple explanation is possible.

Early on, as I probed deeper and deeper into her activities, her thinking, and her writing, it became apparent to me that I was seeing reflections of myself. She saw what I see, but with sharper vision; she heard what I hear, but with keener ears; and her thoughts and beliefs were my own as I would express them if endowed with her ability.

In a sense, getting to know Cordelia Stanwood led me to know myself better, for we have had much in common. Both of us as children had been fascinated by the mystery and the wonder of the woodland. We both became teachers, she for seventeen years, I for eleven; and these careers ended for each of us in a breakdown of health, for her at thirty-nine, for me at thirty-seven. Both of us eventually came to terms with ourselves in the one place where we seemed to belong, a place where the dreams and aspirations of youth acquire substance in the world as it is.

Beyond the Spring was conceived on the night of March 14, 1958 when Agnes Huston and I went to the home of Berenice DeLaite to determine what should be done with the great mass of correspondence, notebooks and manuscripts that Miss Stanwood had presented to our bird club in return for the honor of having it named after her. The field notebooks alone, started in 1905 and continued without interruptions until 1953, would have been enough to arouse excited interest. For they revealed a woman who had devoted much of her life to intense observation and sympathetic interpretation of nature. Agnes, the founder of the club, and Bee, another charter member, knew Cordie Stanwood intimately, but to me she was a shadowy enigma, a rather strange lady who had lived on the outskirts of the community and spent most of her time tramping around the countryside looking for birds.

As the evening went on and the three of us dug deeper into the amazing richness of what was spread out before us, Cordelia Stanwood came alive for me, and I was convinced that I would never be satisfied until I knew everything about her that could be known. The idea that this treasury might get tucked away on the back shelves of a library to gather dust and be forgotten was intolerable. I was committed.

I could see that her life and mine had been related in some strange fashion for a long time, almost as though we had met in the past somewhere but failed to recognize each other.

Some time after I began serious research into Miss Stanwood's background a lost childhood memory came back to me. I couldn't have been more than six or seven years old. A neighbor had loaned my mother a magazine with an article by a woman who devoted practically all her time to the observation and study of bird life about her home. My mother, a teacher prior to her marriage, was much impressed by the story and announced with ill-concealed regret that she too might have done something like this if she hadn't had children and a household to look after. I now recall having felt somehow responsible for preventing my mother from doing something she very much wanted to do, but hazily confused about just what it was I had done. The author was, of course, Cordelia J. Stanwood, and this article was one of the first she wrote. I don't even remember what bird had been the subject of that particular story.

Some twelve years later, in 1925, when the Stanwood home was still pretty much "out in the country," I was knocked unconscious in a car accident on Beckwith Hill just below her house. The car in which I was a passenger had struck and killed a horse. Cordie would have heard the crash and been sorry for the horse, but oblivious to the fact that her future biographer was in some trouble only a few feet from her front door.

Later, in the 50's, I saw once the sad spectacle of an old lady trudging along the shoulder of the road on Ellsworth's High Street, holding fast to a flapping umbrella while she struggled against the March wind and was splashed by the cold wet slush thrown up by passing cars. For me, in this one brief moment, time and change stood still to let this woman pass by with her memories.

On another day when making one of my routine calls at Alice Lord's on Water Street (I was Alice's insurance agent) I became aware of a still, silent figure in the corner of the kitchen. There was the usual lively conversation with Alice, a large, friendly person, but neither word nor motion from that quiet presence in the background. I knew her name, I knew where she lived, but that is all I knew. Who was she really? What would she say if she spoke or put words to paper?

Yes, Cordie Stanwood and I crossed paths in various strange ways, but we were strangers. Our paths crossed but did not converge; our lives touched but did not merge. Only once did I ever meet her, only once did we ever talk to each other--in March of 1958 barely eight months prior to her death, when I called on her in the nursing home where she spent her last three years. My desire for that meeting clouded my judgment since I failed to realize that a woman ninety-three years old afflicted with a terminal illness would be withdrawn within her memories as she approached her last birthday. If I wanted her to answer my many questions about her life I had waited too long.

The absence of any real personal relationship between Miss Stanwood and myself has actually proved to be an advantage. With a minimum of personal preconceptions I have been free to portray her life as the rich materials she left behind reveal it.

Cordelia Johnson Stanwood was a reflection of her time and place--that age of Victorian restraint and Puritanical righteousness which perpetuated the New England Conscience all the way from the fishing hamlets of Connecticut to the rocky hills of Maine. She grew up in an atmosphere in which women were expected to become ladies, with little or no thought as to who or what they might become as persons. Cordelia herself was quite definitely one of these ladies. But, admitting that Cordelia Stanwood was molded by both her ancestry and heritage, we can recognize that she was much more. She was a rebellious soul, a perfectionist and idealist who never entirely abandoned the illusions of her youth, a dreamer forever willing to take that extra step in the eternal search for identity. When the ways of the world failed to offer an answer to her passionate quest, she left the common road, defied convention, and searched with tireless patience to find fulfillment through the intricate balance and mystery of nature. It was here that she found herself, achieved self-expression, and eventually established that harmony of spirit and belief immune to worldly threat. Few, indeed, ever attain this victory over the ever-present frustrations and disappointments of life. In what she sought so tempestuously and in what she ultimately accomplished as a pioneer ornithologist, we can see reflections of our own dreams, hopes, and aspirations through which we either succeed or fail according to our persistence and integrity.

On the upper slope of a hill just south of Ellsworth, Maine, Captain Roswell Leland Stanwood built in 1850 a home for his mother, his two older sisters and their families. Fourteen years later, his bride, Margaret Bown Stanwood of Cape Breton Island, Canada, became mistress of this white Cape Cod cottage with latticed green blinds looking toward the Dedham Hills fifteen miles away on the stagecoach road to Bangor.

Here, in 1865, Cordelia was born and here she spent her early childhood. She was the first of five children born to Captain Stanwood and Margaret Susan. Maude, the most brilliant of the four girls, contracted tuberculosis and died when only eighteen; Maria was married, had one son, and later became a successful lawyer in Boston; Idella, the one to whom everyone turned in times of trouble, was a capable school teacher and married a doctor; and Henry B., the last child and only son, known as "Big Chief" Stanwood, operated a hunting lodge at Tunk Lake and for many years was fly-casting champion of Maine. At the age of fourteen in 1879, the year her brother was born, Cordelia was sent to Providence, Rhode Island, to live with the wealthy aunt for whom she was named, so that she might enjoy a broader education than Ellsworth could provide. She was graduated from the Girls' High School of Providence in 1886, ranking sixth in a class of sixty. She attended Normal School and later the Normal Arts School in Boston, and for seventeen years was an effective and dedicated teacher and administrator in Rhode Island, Massachusetts, and New York. At one time she taught in a training school for Vassar College.

Forever seeking something just beyond whatever she had achieved, she became increasingly impatient with the restraints of her profession and the regulated narrowness of the school teacher's role in society. Her rebellion against frustration led to a breakdown in health which forced her to give up her teaching career and return to her home in Maine. It was the turning point in her life, the challenge which could not be evaded if she was ever to justify her ideals and truly function in accordance with her capabilities. The character of the woman is revealed in her triumph. She overcame broken health, chose her objective, and achieved familiarity and mastery of woodland and field. Cordie Stanwood truly found the richness of life in the hidden, quiet places, and she generously shared this richness with all who were willing to observe, to listen, and to respond as she had done over the long, full years.

The book of her life closed on November 20, 1958. She was ninety-three years old. Only a few people were present at the funeral, possibly twenty. Some were the friends of recent years, others had known her as children, and there were the neighbors. Only the one remaining sister, Della, could remember back to the beginning almost a century ago. It was for these few to offer silent tribute to a gifted, indomitable spirit who had brought her vision to fulfillment.

Eagerly, I have reopened that book closed by the finality of death in 1958; for there is still much to tell. That which does not die, that which lives on in the minds and hearts of others speaks out in the pages that follow. During her lifetime few listened, too many had not heard.

Listen to what she had to say and meet the real Cordelia Stanwood.

1

Looking Back

Thanks in old age . . .
For precious ever--lingering memories.
Whitman

The clapboarded sides of the little house on the hill overlooking Ellsworth were weatherworn and gray from years of neglect. Only a few faded shades were visible through the small-paned windows with no curtains to soften the suggestion of empty staring eyes. Tall matted grass choked the dooryard and all but concealed the sunken granite step before the doorway into the ell. The lilac and japonica were ragged and forlorn remnants of the former splendor that once vied with the apple blossoms beyond the barely visible outline of the driveway.

Sitting on a camp stool and enjoying the warmth of the Indian summer sunshine was an old woman. She was so much a part of the drab house and overgrown dooryard that one might easily have passed by quite unaware of her presence. As she leaned forward, thin blue-veined hands smoothed the wrinkles in the long black skirt, and a smile crossed her face as she watched a downy woodpecker hammering a balm-of-gilead stump nearby.

She was alone, yet not alone. Her whole world was before her--the fields, the hills, the forest, the shadbush and the arbutus, the wild cherry and the linnea, the flight and song of birds. This was the world she had always known and loved, the world that had brought meaning and purpose into her life when all else failed.

Now it was 1954, but her thoughts carried her back seventy-five years to a Sunday afternoon in August when she had walked down to the Boiling Spring. Her fourteenth birthday! Golden slanting rays of sunlight had filtered through the

Cordelia Stanwood at fourteen, her first year in Providence.

branches of the twisted pine across the hollow as a dusky bird fluttered upward from the mossy bank into the canopy above spraying liquid diamonds into the still warm air. She could still hear the clear sweet ripple of its song.

When a cloud blotted out the warm sunshine, she shivered and pulled the faded gray sweater closer around her shoulders. Then she arose, picked up the camp stool, closed the door, and slowly climbed the stairway leading to the ell chamber. The disorder of the room above, the dingy smoke-blackened walls, the unmade bed, went unnoticed. She was still with that girl at the spring and hearing the song of the thrush.

Why had she gone to the Boiling Spring that day? What led to that tingling excitement and breathless expectancy she had felt so keenly when she ran up the cowpath to the pasture bars? Such a happy memory. She had been about to set out on a long trip--by stagecoach, boat, trolley car, and train. Within a week she would be on her way to Providence, Rhode Island, to live with Aunt Cordelia and go to school. The beautiful new dress of soft white linen with the dainty blue design was finished and hanging in the closet, and her trunk was nearly packed. Never again would she know a time so filled with anticipation and joyous dread as that summer of 1879. She wrote in her journal: "I would have crawled on my hands and knees to go to school in Providence."

Yet Cordelia Stanwood was never one to crawl, physically or mentally. In truth, she sometimes held her head too high and got some hard bumps. Widely known as the Lady of Birdsacre, this old woman remembering her childhood would be ninety years old on her next birthday, August 1, 1955, still alert and filled with zest for living and curiosity about the world around her. For half a century this alertness and curiosity had drawn her intimately into her natural environment and by sharing this experience with others through her photography and writing she had earned her place among naturalists.

Now that she was old and rather tired she no longer tramped through the woods searching for better understanding of bird life. But this withdrawal neither dampened her enthusiasm nor clouded the vividness of her memories. She saw in recollection the spread of green grass beyond the drive and the mist of pink and white blossoms on the apple trees. She heard the jingle of sleigh bells as sturdy horses pulled grain-laden sleds up the hill, and the ring of the axe against a tree trunk on a frosty morning when the winter's firewood was cut. The fragrance of spruce and balsam brush was still strong in the air as she listened to a bird song from the lilac bush, and she danced with the yellow roses swaying by the doorstep. In the background was the click and flash of Grandma Stanwood's knitting needles in the lamplight, her mother's soft voice and her father's sternly quiet presence. She could still hear the patter of raindrops on the window-pane in the comforting darkness of the night while she lay snugly beneath her patchwork quilt.

How happy I was, she thought, as she placed the dented teakettle on top of the box-like woodstove for her afternoon cup of tea.

She recalled her father's bearded face behind the helm of his ship when she and her mother had been with him on one of his many voyages along the Atlantic seaboard. Once, after a terrifying storm off Cape Hatteras, it had taken thirty days for the crippled water-logged ship to flounder through heavy seas to the safety of New York Harbor. She saw again the glint of the small gold coins she had played with while they had been confined below decks in the narrow cabin. That had been the last voyage for her mother, who was not a good sailor.

Less dramatic but a happier memory was a sailboat trip across Patten Bay at the mouth of Union River, followed by an overnight stay on Newbury Neck with a friendly farmer's wife called Aunt Jane. Her father always had his own small boat to keep him close to salt water when he was home from the sea. Cordie's sisters Maria and Idella and two neighbor children were also in the party. At the mouth of the river Father caught some cunners, then landed at Shady Nook to prepare a chowder. After securing some milk at a nearby farmhouse, he cleaned the fish, built a driftwood fire, and created a savory meal for five girls very hungry after a romp on the sandy beach. Later a brisk breeze sent the little boat skimming across the bay where supper at Aunt Jane's was a banquet of fricasseed chicken, butter biscuits, wild strawberry preserves and cake. Filled with good food, salt air, and sunshine the girls crept under the patchwork quilts in the special bed made up by Aunt Jane on the bedroom floor and knew no more until the Captain routed them out early the next morning for the return trip. The lively breeze that sent them scudding across Patten Bay seemed a near hurricane to Mother until they were safely home again.

Such experiences tumbled pleasantly through her mind as she ate her crackers and cheese and sipped her tea. It was good to remember. And this present lonely way of life so deplored by friends and neighbors was no loneliness at all.

Life was much too full for that. It had ever been so.

Happiest when pursuing her studies, Cordie Stanwood was an insatiable student throughout her life. When she went to Providence in 1879, her eagerness to learn made her impatient with every delay which kept her from the classroom. For the next seven years she gloried in the broad cultural advantages of a large city school system, and she was forever grateful to Aunt Cordelia for making it possible.

But it had not been easy. Rather frail as a child, she had not attended school regularly until after her eighth birthday. As a tot she had sat on a stool at her Grandmother Stanwood's knee and learned to knit, darn, and make patchwork, and her mother taught her to read, but at best this early schooling had been sketchy. She remembered how painfully self-conscious she had felt to be older than her classmates. Sensitive and diffident, she could not make friends easily. This inherent shyness caused her to live pretty much in a world of her own.

In the rich splendor of her Aunt's Broadway home she was among elderly people unaccustomed to children. She studied with boys and girls who could not possibly understand the life pattern of a girl from a black-forested hillside in Maine. Consequently she held herself aloof, so that to most people she would always seem unapproachable. This tendency to reserve became a permanent part of her personality and led to that solitary approach to life so compatible with her long involvement with wild nature.

On many a day while living in Providence she had slipped away to the Victorian elegance of the front parlor to sit in one of the crimson embossed satin-covered chairs, to see her reflection in one of the full-length mirrors, and to wonder about it all—who she was, where she was going, and what she would become. She loved the soft gray of the imported Brussels carpets, the Duncan Phyfe bookcase and matching writing desk, and the heavy draperies across the double parlor doors. That pensive girl who gazed back at her from the tall mirror was so intense, and the questioning blue eyes in the angular face with the wide mouth and prominent Stanwood nose demanded answers. Here by herself in the parlor she could cope with her shyness and renew her courage.

It was a help to remember what her Aunt told her many

times:

"Cordie," she would say, peering over her gold-rimmed glasses, "never forget that you are a Stanwood, that your great-great-grandfather Job lost an arm at the Battle of Louisburg during the French and Indian Wars, that other Stanwoods fought in the armies of Washington. They were a proud, sturdy lot; so keep your chin up and your spirit high so that you will be worthy of your name."

Cordelia Stanwood in 1882, the year in which she entered high school and became a member of the First Baptist Church.

Aunt Cordelia, generous and aggressive, never wearied in her efforts to help others improve themselves, and she had been especially zealous for her namesake. Even in the matter of clothes, she made certain that Cordie's dresses were becoming and well-fitted by taking her to her own dressmaker. Whatever had happened, Cordelia wondered, to that peacock-blue dress and the beautiful hat with the ostrich plume which she had worn on that first visit to the First Baptist Church founded by Roger Williams. How thrilled and awed she had been to sit in the Johnson pew as one of the family. As they had walked home down Broadway, the colorful brightness of her own appearance had contrasted sharply with the somber black of her Aunt's dress and Uncle Oliver's tall black hat and white beard.

And so it had been with her through her girlhood, those fortunate years full of promise and challenge. Diligently, methodically, patiently, she had adapted and absorbed.

In June of 1886 she had stood proudly in sixth place in a girls' class of sixty. Twenty-one, slender and straight-backed, she reached forth eagerly to open the door into the future. Almost imperceptibly she had absorbed the gracious genteel way of life so much a part of the Johnson home. To hear and participate in good conversation, to know and mingle with broadly experienced and refined people, to learn how to be at ease, courteous, and well-spoken, could hardly fail to inject into her character that dignity of self-assurance and confidence she needed to protect that proud and sensitive other Cordelia who could never quite allow the outside world to reach her.

Cordelia set very high standards for herself very early in life, standards which she would cling to through every challenge. Living on Broadway with one of the most socially prominent families of Providence provided her an atmosphere for growth and refinement. It did not make her a snob, it created a lady.

Uncle Oliver was a deacon of the First Baptist Church. Cordie attended the services in this church, went to Sunday School classes taught by Brown University professors, and became a member when she was seventeen years old. As late as 1932, long after she had returned to Maine, she still sent her modest contribution to her church in Providence. In the home of Aunt Cordelia and Uncle Oliver Johnson Cordie absorbed the social and religious values of the church of Roger Williams. Anything less than the highest standards of personal conduct would always be unacceptable to her, and at no point in her long life would she ever falter in her commitment to those standards.

2

A Family Heritage

Think of your forefathers!
Think of your family name!
John Quincy Adams

Cordie's life-long interest in family history began when she was a young girl, even before she went to Providence to live with Aunt Cordelia. She loved to listen to Grandmother Stanwood tell about the days when the family lived in Eden (now Bar Harbor) on Mt. Desert Island. When she was only seven, she heard many anecdotes about her ancestors while her grandmother taught her to knit mittens and stockings, and to piece a bedquilt. Years later Cordie recorded in her notebooks her memories of what she had learned from her Grandmother Stanwood.

For example, the early settlers on Mt. Desert Island were frequently cut off from the mainland and supplies for many weeks at a time. When asked if they ever suffered from lack of food Grandmother Stanwood promptly replied: "Oh no, the cellars were always full of vegetables, beef, mutton, fish, eggs and the like." And this would lead to more and more detail about what went on in those times:

> *They killed a calf and tanned the hide when anyone needed a pair of shoes. There was a cobbler in the community, if not in every home. The families were in the habit of shearing the sheep, spinning the wool into knitting yarn, or into warp and woof to use in weaving. White homespun cloth was sent over to the mill between the hills in Somesville to be fulled and dyed a bright scarlet.*

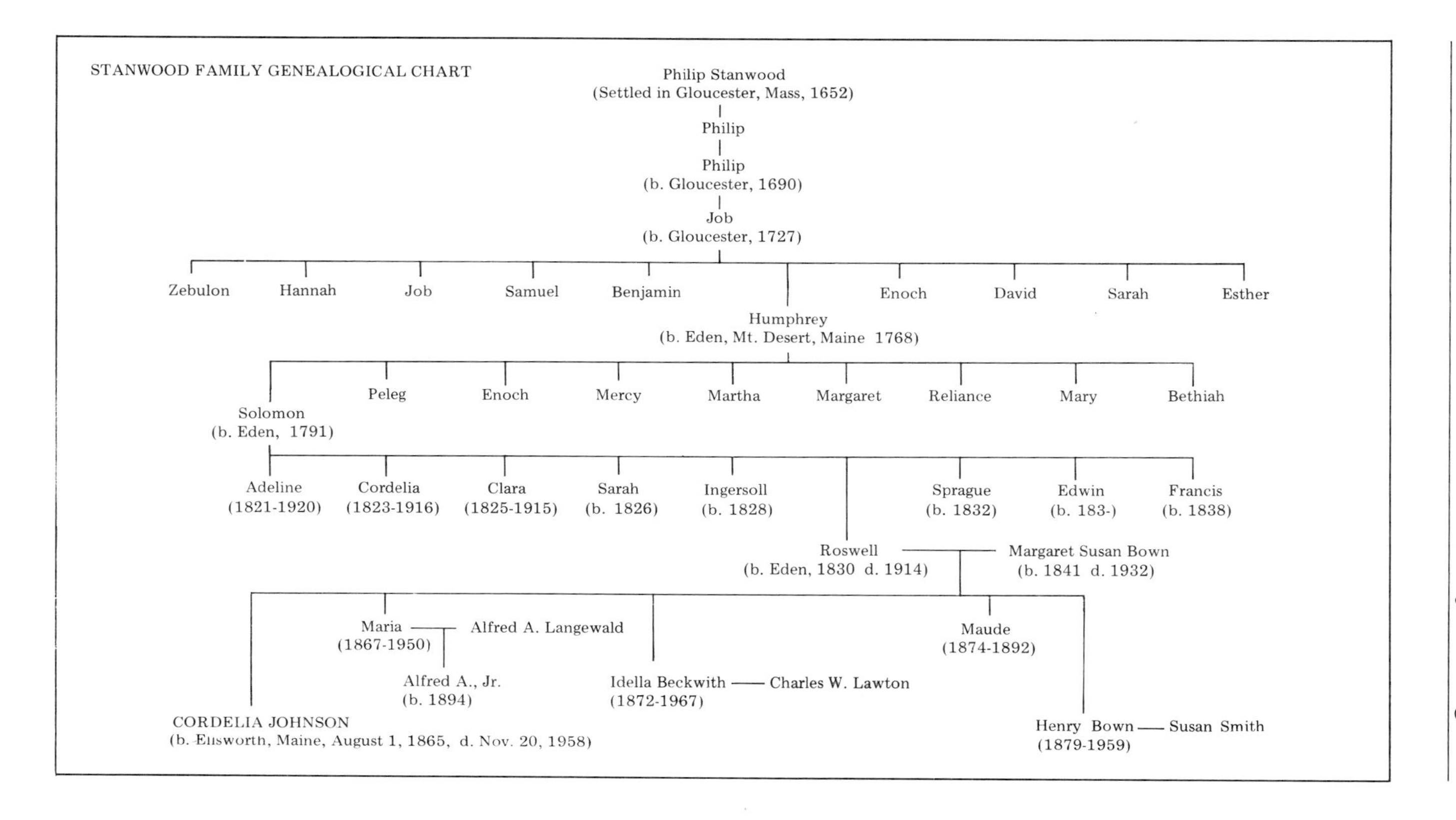
STANWOOD FAMILY GENEALOGICAL CHART
Philip Stanwood
(Settled in Gloucester, Mass, 1652)
Philip
Philip
(b. Gloucester, 1690)
Job
(b. Gloucester, 1727)
Zebulon
Hannah
Job
Samuel
Benjamin
Humphrey
(b. Eden, Mt. Desert, Maine 1768)
Enoch
David
Sarah
Esther
Solomon
(b. Eden, 1791)
Peleg
Enoch
Mercy
Martha
Margaret
Reliance
Mary
Bethiah
Adeline
(1821-1920)
Cordelia
(1823-1916)
Clara
(1825-1915)
Sarah
(b. 1826)
Ingersoll
(b. 1828)
Sprague
(b. 1832)
Edwin
(b. 183-)
Francis
(b. 1838)
Roswell
(b. Eden, 1830 d. 1914)
Margaret Susan Bown
(b. 1841 d. 1932)
Maria
(1867-1950)
Alfred A. Langewald
Maude
(1874-1892)
Alfred A., Jr.
(b. 1894)
Idella Beckwith
(1872-1967)
Charles W. Lawton
CORDELIA JOHNSON
(b. Ellsworth, Maine, August 1, 1865, d. Nov. 20, 1958)
Henry Bown
(1879-1959)
Susan Smith

The gowns worn by the children were woven at home from the wool cut from the backs of their own sheep. The girls were not only excellent knitters and weavers, but fine needleworkers. The tucking, ruffling, and the like which they did were as beautiful as one can imagine. It was almost impossible to see the stitches. The women wore homemade bonnets instead of hats.

This habit of meticulously recording everything which her grandmother told her about her forebears prepared Cordie for the methodical research that eventually made her the authority on the Stanwoods of Maine. When Ethel Stanwood Bolton published the first definitive genealogy of the Stanwoods in 1899 it was Cordie's research which accurately and precisely placed her own branch of the family among the many who had come from that first Philip Stanwood who had settled in Gloucester, Massachusetts, in 1652.

She was proudly aware of her father's long and colorful career as a seaman. Roswell Leland Stanwood had gone aboard his first sailing vessel as an eleven-year-old cabin boy and when only twenty was master of his own ship. An independent and resourceful captain he had "run the gauntlet" of the blockade during the Civil War, steering a straight and steady course through hazards which discouraged less courageous mariners.

In 1881, when she was sixteen, Cordie wrote a biographical study of Uncle Oliver Johnson, showing her ability to dig out facts and assemble them in logical and interesting sequence. Uncle Oliver, then in his eighties, was a pillar of respectability in the business world and community affairs of Providence, and she saw in him the same qualities of character and personality she had come to admire in her own family background, a background rich in the traditions of an earlier America.

In the summer of 1890, just prior to her twenty-fifth birthday, Cordelia brought to fruition a dream which had persisted since her graduation from high school—to go to Canada and visit her mother's parents and look upon the scenes of historical events that dramatically linked her father's and mother's families back to the French and Indian Wars of the eighteenth century. Before this visit her only image of New Brunswick, Nova Scotia, and Cape Breton

Island had been of a solid-color land mass lying to the north of New England in a geography book. When she finally set foot on this vast sprawling land she was pleasantly amazed to discover how closely it resembled the landscape of her home in Maine.

It was a joyous vacation and an exciting adventure to travel so far north. Not only was Cordie's enthusiasm increased by virtue of her fascination with family history, but she also brought healthy high spirits and all her senses open to the richness of the experience. At St. John she had been invited by Professor and Mrs. Chickering and their daughters from Washington to join in an excursion on the St. John River. They saw the reversible falls and had at Hamstead a salmon trout dinner with wild strawberry dessert. In her journal she carefully recorded her observations:

> *I think the scene was what the poet called "silence rendered visible." The river was very broad and smooth as glass. Occasionally a huge raft of immense logs sailed noiselessly by and in the distance one could hear the tinkle of cowbells. The air was fragrant with red clover and wild strawberries.*

On the streets of St. John she felt a strong impulse to stop the people whom she met and inform them that this bustling community, their home, had been largely settled through the efforts of her great-great-grandfather, George Leonard, who, as a loyal British subject under King George III, had been most responsible for settling Loyalists here during the Revolutionary War. But Cordie was not one to indulge impulses.

At Eskasoni on the shore of East Bay of the great Bras d'Or Lakes which dominate the interior of Cape Breton, Cordelia was warmly welcomed by grandparents who had celebrated their golden wedding anniversary in 1884, six years before. This visit had special significance for them since Cordie was the only one of Margaret Susan's five children they had the opportunity to know. Cordie, in turn, was thrilled to find herself at Eskasoni, where three roots of her own ancestry linked together. Within a radius of less than fifty miles here at Cape Breton were the forces of family and history that had attracted Bowns, Leonards, and Stanwoods to this most eastern outpost of North America and bound them together in marriage. In her grandparents' home at Eskasoni, she was not just visiting her mother's parents, she was absorbing events of history and responding to the drama and romance which had brought her father, together with the dynamic spirits of both great-great grandfathers, under this one roof. She felt both proud and humbly grateful to be there.

Her maternal grandfather, Henry Vincent Bown, came from a family of Huguenots who had fled from persecution in France and emigrated to Canada to develop the coal mining industry. A resourceful and conservative family, the Bowns had been leaders in mercantile activity in the port city of Sydney since the late eighteenth century.

When still quite young Henry fell in love with and married Maria Jane Leonard, the aristocratic granddaughter of George Leonard, who had figured prominently in settling Loyalists at St. John, New Brunswick. Henry relished every opportunity to relate that his charming wife had lived with her illustrious grandfather at the family estate, Sussex Vale in New Brunswick, and in 1859 when only eighteen had been privileged to lead the grand march at a state function with the Prince of Wales.

Since Cordelia had already painstakingly catalogued the impressive lineage of the Leonard family back to a castle in Sussex, England, and beyond that to two brothers who had crossed the English Channel with William the Conqueror in 1066, she became quite excited when her grandmother brought forth the copy of a St. John newspaper story written in 1826 recounting George Leonard's accomplishments during and after the American Revolution. She was amazed to read that he had been completely involved in those long eight years of conflict and sacrifice, never faltering in loyalty to his king and the cause of England.

> *In 1775 Mr. Leonard was a volunteer under Lord Percy and was engaged at Lexington, the first battle fought with the Americans. He assisted as Lieutenant under Col. Willard with the British troops in the defense of Boston in the same year . . . . In the year 1776 he left Boston with the British army for Halifax and went from there to New York and was at the taking of that city. . . .*
>
> *He purchased, fitted for sea, and manned at his*

own expense a small fleet, consisting of seven armed vessels and three transports, besides several gunboats, for the purpose of harassing the enemy on the seacoast and cooperating with his Majesty's army in making descents upon the coast whenever called upon by the British Admirals or Generals. . . .

At the evacuation of New York by the British Army in 1783, he was appointed one of the agents to settle the Loyalists on the Crown Lands of Nova Scotia, and the establishment of them at the mouth of the River St. John, now the city of that name, fell to his lot.

George Leonard's career of service to his fellow Canadians continued for many years thereafter until his well-deserved retirement to his beloved country estate, Sussex Vale in New Brunswick, where Maria Jane had spent several happy years prior to her marriage to Henry Vincent Bown.

Cordie had been at Eskasoni only a few days when she went with her cousins in Sydney to visit the ruins of the Fortress of Louisburg, at one time the bastion of French influence in the western world, later to be called the "Dunkirk of America." Standing on the ruin of one of the great walls and looking out across the land-locked harbor toward Lighthouse Point, she was very conscious that another Stanwood had been here before her. Among that motley crew of Colonials who had sailed up the Atlantic seaboard in Colonel William Pepperell's armada in 1745 to launch the siege of Louisburg was Job Stanwood, great-grandson of the original Philip. Job, only eighteen at the time, had lost his left arm in that bitter struggle, but more significantly he had been struck by the grandeur and beauty of the Maine coastline as seen from the deck of his ship as it sailed by on its way to Cape Breton. Permanently handicapped by his participation in the Louisburg venture, Job returned to Gloucester and after his second marriage, to Martha Bradstreet, a minister's daughter, moved his family to the Cranberry Isles which faced the mountains of Mt. Desert Island. This had been around 1760. Later, he settled in Eden, now Bar Harbor, on Mt. Desert.

Job's children were the forerunners of all Stanwoods east of the Kennebec in Maine, and his second son, Humphrey, was Cordie's great-grandfather. One of Humphrey's brothers, Enoch, had moved to Yarmouth in the Provinces as a young man, was married there, and remained loyal to the king. During the last war with England in 1812 he commanded a privateer which took prizes off the coast of Maine, and he finally lost his own life in a scrimmage near Deer Isle.

The dream of French empire in the western hemisphere had died at Louisburg, and Canada remained loyal to England, but America gave birth to a young nation committed to a new concept of freedom and self-government, that same kind of freedom and self-determination Cordelia Stanwood demanded from life. In that great struggle the Stanwoods, unlike the Bowns and the Leonards, were more inclined to cast their lot with the cause of liberty and independence.

One of them, and probably the most prominent, was Colonel William Stanwood who, at one time, was on General Washington's staff in Pennsylvania. Between 1796 and 1798 this same Stanwood "sold to the Trustees of Bowdoin College, for one cent" land that "comprises what is now the campus" of that institution.

Another Stanwood from the Augusta branch of the family became the wife of presidential aspirant James G. Blaine and maintained an estate on Mt. Desert Island which she called "Stanwood." Cordie took pictures of this place before it was destroyed in the Bar Harbor fire of 1947.

Cordelia was fascinated by the interplay of her family with history. In the process of creating a new nation her ancestors had been directly and intimately involved as activists on one side or the other, and with equal vehemence on both sides. At Eskasoni she felt a part of it as a related observer from a later generation, free to express her sympathy and understanding for both sides.

A few days before she reluctantly said goodbye to her grandparents to return to her Providence schoolroom, Cordelia slipped quietly into the parlor of the Eskasoni homestead to sit by herself and sort out the many emotionally exciting experiences and discoveries she had found during the happy days she had been there. She was not intimidated by the somber severity of this seldom-used room, for it was here that her father and mother had been united in marriage on Margaret Susan's twenty-third birthday, September 14, 1864. Sitting here she rehearsed in her mind the story of their meeting.

In the summer of 1863 Captain Stanwood had docked his vessel in the harbor of Sydney, Cape Breton Island, to take on a cargo of coal. It was a routine voyage for him, and it is doubtful that he even remembered that he was within a few miles of Louisburg where his great-grandfather had been severely wounded over one hundred years before. He was there on business, to secure a profitable load of coal to be delivered to a customer in New York.

Captain Roswell Leland Stanwood of Ellsworth, Maine, father of Cordelia Stanwood.

Thirty-two, and the owner of his own ship, he sought out the coal mining interests with confident assurance and negotiated for his cargo. After concluding a friendly and mutually satisfactory understanding in their business relationship, Henry Vincent Bown of the coal company had invited

Roswell to visit his family at Eskasoni. He accepted the invitation rather casually, but was immediately impressed by the warmth and friendliness of the Bown family and quickly realized that he was strongly attracted to their daughter, Margaret Susan.

Margaret, however, stood in awe of this tall sea captain, her senior by ten years, but she soon lost her shyness after several visits before he had to sail for New York.

Margaret Susan Bown Stanwood, mother of Cordelia Stanwood.

Mr. and Mrs. Bown saw in Captain Stanwood an acceptable suitor for their daughter's hand in marriage, but in keeping with the times and their own backgrounds they withheld final approval until he had produced references as to character, background, and reliability.

So, here on Cape Breton Island almost in the shadow of the Fortress of Louisburg, the Bowns, the Leonards, and the Stanwoods would now be permanently linked through the marriage of Margaret Susan Bown and Captain Roswell Leland Stanwood of Ellsworth.

Sitting in the parlor where they had been married, Cordie recreated the scene in her imagination. The melodeon against the west wall by the window facing the Bay had been played at the wedding. One day Grandmother Bown had played it for her so that she could hear its mellow tones. She took pleasure in visualizing the simple ceremony through which the Bowns, the Leonards, and the Stanwoods had become merged in that common heritage which gave substance and meaning to her own life. She would not exist except for them, and most certainly she would not be who she was without that strong mixture of blood line and cultural background which had brought them together.

It is not difficult to understand why Cordelia was so proud of the name she bore. It committed her to a standard of excellence that she would apply more rigorously to herself than to others. This pride would drive her to her own personal distinction and would sustain her in the loneliness of her later years.

3

Teaching Days

Wait, thou child of hope, for time shall teach thee all things.
Tupper

One April morning while rain beat a steady tattoo on the ell roof shingles, Cordelia unlocked the heavy trunk which held the story of her teaching days. Crammed full of textbooks, records, lesson plans, magazines, pictures, and sketches, this trunk scattered the dust of time, brought back the light and sound of classrooms from Providence to Poughkeepsie and surrounded her with the faces and voices of numberless children.

How remote it had become. With a rueful smile she reminded herself that an old lady should certainly feel no compulsion to prove anything to herself or anyone else. She had outlived the need to compete with others in order to establish her own worth or to achieve some goal. What pleasure just to indulge in remembrance without rancor or envy, to recall the good and ignore the unpleasant, to remember that, after all, she had been a good teacher and that she might have become an even better one if health had not forced her away. Had she not accomplished more, possibly, through her photography and writing as a naturalist than she ever could in the confinement of the classroom?

Those seventeen years of teaching really belonged to another life. They had, however, helped to shape the character and personality of the Cordelia Stanwood who became known as the Lady of Birdsacre, and had not hampered that freedom of thought and action so essential in everything she did. Her reputation as a teacher had been exemplary. With the solid background of a classical education in the excellent Providence school system, supplemented by the Teachers' Training School, the Boston Normal Arts School, and courses at Harvard and elsewhere, she was qualified to take her place with the best.

Cordelia Stanwood in 1887, the year she began her teaching career.

Memories of childhood, of school years, and then the teaching: years of challenge and excitement, occasionally boredom and frustration when she had been unable to find sufficient outlet for the boundless enthusiasm and energy she applied to every venture. She told herself that she must then have been a very impatient soul, for there had been nine different schools in those seventeen years, and she had never taught more than two years in any one of them. Those schoolrooms had reached from Providence, where she started with a first grade in 1887, to Springfield, Greenfield, Brookline, Braintree, Marion, and Plymouth in Massachusetts, and to the Quincy School in Poughkeepsie, New York. Somehow she had never been able to understand her restlessness and rebellion until circumstances forced her to abandon her teaching and return to Maine. How fortunate, she told herself, to have found her way in woodland and field after illness threatened everything she had assumed to be essential. None of it had been that essential, none of it that important. Now, looking back, she realized how thoroughly she had been dominated by a drive to get somewhere, when in reality all she honestly wanted was to be herself.

Greenfield, one of the happiest of her teaching experiences, illustrated this well. She had been supervisor of drawing and art work in a group of twenty-nine schools surrounding the area, a situation which placed great demands on both energy and time. She had indeed been successful, but her teaching success was not what caused her to recall those days with most pleasure. It was the memory of the spontaneous happiness and freedom she had felt so keenly when away from the classroom and travelling about the countryside from school to school. Tucked among the western hills of the Bay State, Greenfield lies at the junction of three rivers, the Connecticut, Deerfield, and Green. In the Nineties it was still pretty much a rural community where many of the homes were owned by descendants of those first settlers who had received their original land grants from King George III. In this natural garden spot Cordie had found rich opportunity to become better acquainted with the wild flowers and bird life of the valley.

Quite frequently Superintendent Dame, the Reverend Dr. Finch, or Miss Mims, a member of the school board and a former teacher of botany, accompanied her on these trips to outlying schools.

> *With Miss Mims I became acquainted with many new and rare flowers. We found the pale blue crowfoot violet growing under the trees on the wide grassy sides of the street in Deerfield, and the yellow violet among the leaves in the woodland. The elusive maiden-hair fern flourished about the damp rocks overhanging the turbulent Green River, and gay bittersweet vines climbed decaying stumps along old stone walls. At the watering*

place I saw the scouring rushes used by the early colonists to clean their milkpans.

I think it was Mr. Dame who pointed out patches of golden cowslips blossoming in the little streams that threaded the meadows, and at laurel time the pink and white blooms which covered the banks of the river, sometimes to a height of ten feet, were almost breathtaking in their beauty.

In a beautiful bit of woods here in Greenfield I heard my friend, the black-throated green warbler, for the first time--called to my attention by a teacher companion who had studied birds. It seems strangely wonderful to me that she could identify him through his song alone.

Cordie listened to the steady patter of the rainfall on the roof just over her head as she sat in the ell chamber with her memories. Reaching forward she picked up a Christmas card tucked in among the school papers, a card from Dr. Henry Turner Bailey showing one of his pen sketches of a church steeple near his home in North Scituate, Massachusetts, a place she had visited on several occasions. Each year since 1897 she had looked forward to these artistic greetings created by Dr. Bailey for his many friends. The last one had come in 1930 only weeks before that fall on the icy streets of Cleveland which had led to his death. How she had missed their correspondence and the steady encouragement he had generously provided.

She no longer heard the rain and was hardly aware of where she was as her mind carried her back to those days of long ago. So much of what she had accomplished had come about through her association with Henry Turner Bailey--author, artist, educator, and lecturer--a man who had inspired students, teachers, and artists from Massachusetts to Ohio to their greatest achievements, a man of about her own age, handsome, gracious, and married.

Looking back it seemed almost incredible to her that she had been so reluctant when Miss White, the art instructor in the Providence school system, had urged her to hear her first lecture by this dedicated man. This was in 1890 when she was twenty-five and teaching at the Eddy Street School. Previously Cordie had avoided such programs since most of them were extremely dull. Dr. Bailey, however, proved to be surprisingly different. She had been fascinated by his tremendous enthusiasm and sincerity; he made her feel personally involved in everything he had to say.

One short hour in the auditorium of the high school that night marked the beginning of a relationship that would change the rest of her teaching career and profoundly influ-

Dr. Henry Turner Bailey, artist, author, teacher, and mentor to Cordelia Stanwood. This picture, clipped from a magazine, shows him at about the time when they first met, in 1890.

ence the direction she would take thereafter. She even recalled how much his sympathetic and constructive criticism had helped her develop a style of writing that made her bird stories attractive to publishers.

In the following summer of 1891 she had gone to the summer school at Martha's Vineyard to register in one of his art courses, and she studied under him again the next year at Glens Falls, New York. So when Dr. Bailey recommended that she seriously consider becoming a full-time art teacher, she gave up her principalship at the Plain Street School in Providence to enter the Normal Arts School in Boston. The Providence years were over.

At twenty-nine she was eagerly confident as she welcomed the challenge of those years ahead. More than thirty years later in 1926 when he was Director of The Cleveland School of Art, Dr. Bailey recalled of Cordie:

Cordie, at about 31, as Supervisor of Drawing in Greenfield, Massachusetts.

I have known her for many years and have found her a persistent worker for self-improvement and for helpfulness to others. I have pleasant and vivid memories of her as a student in the Summer School at Martha's Vineyard, Mass., and Glens Falls, New York, during my service there as instructor, and of the Harvard Summer School where I was a fellow pupil with her in the classes of Dr. Ross. In every case Miss Stanwood's work was of high quality.

Within the closely guarded privacy of her own heart, protected by the immunity of old age and solitude, she could now admit to herself that Dr. Bailey had meant far more to her than she had ever been willing to acknowledge. Challenged by his scholarship, inspired by his artistic insight, and guided by his wisdom, she could hardly confine her feelings to the admiration and respect of strong friendship. Her relationship with Dr. Bailey was the nearest approach to a romantic attachment that Cordie ever made. She never married.

The background influences of a Victorian environment and the upright people with whom she lived and studied through those teenage years in Providence had left their stamp upon her. Uncle Oliver was the epitome of all that a man worthy of respect and admiration should be, and the aunts brought her up in the tradition of their own upbringing. "I can recall," writes her sister Della, "seeing her dressed in white with two braids of dark hair pinned together with a brilliant clasp that sparkled in the sunlight. The aunts kept her groomed within an inch of her life during those school days in Providence."

"Cordie was not," she adds, "what you would call pretty, more the type of face that one would associate with a George Eliot." Pretty? Perhaps not, yet certainly a striking, colorful young lady with a capacity for warmth and affection that was never put to the test. Somewhere along the way she had missed the commitment between a man and a woman which a romantic union would have provided.

There had been eligible young men during the Providence years, usually students at Brown University whom she had met at the First Baptist Church or through her friends William and Bertha Eddy, but whenever one of them showed any special interest the aunts would circumvent her own responsiveness by subtle implication or biased comparison to Uncle Oliver, in their opinion the only kind of man worthy of consideration in marriage. Inevitably she drifted into an attitude wherein an impossibly idealized masculine fiction replaced the normal, understanding companion who might have won her heart.

As her sister Della has said, "Because of the aunts, Cordie saw 'men as trees walking'."

Although at 33 she was not yet a spinster, her mounting restlessness betrayed the unresolved conflict between her normal femininity and the hard core of a demand for the unattainable which would inevitably lead to spinsterhood. Unfortunately, what she sought in male companionship no man could offer. It involved the idealistic blending of Uncle Oliver with Dr. Henry T. Bailey, garnished with a touch of her great-great-grandfathers, Job Stanwood and George Leonard of Sussex Vale, and Dr. Bancroft of the First Baptist Church.

Brother Harry, despite the bitter feud between his sister and himself, stemming from parallels in disposition and unyielding wills, observed a month before his own death: "Cordie should have married. Things might have been different had she done so." Even her mother had remarked wistfully, "I wish Cordie had married." Most people learn to compromise with the world as it is, but Cordie would not, and could not, compromise her convictions. Her ideals were too lofty, her imagination too vivid, her sense of perfection too stern and exacting. She walked alone from choice, not necessity.

Very much aware of all this as she sat quietly among these mementos of the past, she was neither sad nor bitter. Life had been much too generous to her in so many other meaningful ways.

When she came across a small picture of Elias Vail in a corner of the trunk, Cordie smiled to herself. Children were so delightful, and many of her happiest memories were linked one way or another with their spontaneous enthusiasm and inexhaustible vitality. Elias, one of the pupils at the Quincy School in Poughkeepsie, had taken her for a ride in his pony cart one day out to the college campus, and rather proud to have his teacher for a passenger had said to her when they returned: "Now Miss Stanwood, no one can say you haven't been to Vassar College."

She had always liked children, and they, in turn, responded to her. She was generally at ease with youngsters, while inclined to be reserved and restrained in the company of adults. To be able to relate to them with sympathy and understanding had made her a better teacher, that kind of teacher who could relish the humor in John Adriance's report on a nature walk she had organized one day. Armed with grape baskets and trowels they had dug up a few anemones and some bluets, and John had proudly summarized his review of their excursion in these words: "I planted all my enemies in the flower garden, but my father said he didn't like that kind of flower, he liked roses."

It was getting late and she was tired, so she closed the old trunk and went out to the stove to stir up the fire and make herself a cup of tea.

At 35, at the turn of the century, as teacher in the Quincy School, Poughkeepsie, N.Y.

She had to admit to herself that even after all these years she still found it disturbing to remember how abruptly her teaching career had come to an end. Time had not dimmed the terrible sense of loss and helplessness she had felt when she came to realize that she would never return to the classroom. Engaged in art work and freed of the restraining influence of the Aunts in Providence she had done her best work between 1894 and 1904, but at a cost dangerously close to destroying everything she had worked so hard to achieve.

Sickness is a fact of life over which one has but limited control. Cordie had known this, and had also known but failed to pay heed to the necessity to protect herself against undue risk. Although she could and did control the physical conditions which surrounded her, she had been extravagantly spendthrift in giving of herself. It had brought her face to face with the specter of failure, a failure she could not ac-

At 39, in about 1904, at the end of her teaching career.

cept. It had never been conceivable to her that anything could ever happen to the pattern of life she had fashioned for herself, and when the first evidence of her breakdown began to manifest itself she had chosen to brush it aside as something quite meaningless like a bad dream. But the bad dream had persisted and quickly grew into a nightmare that had nearly destroyed her self-confidence and self-assurance and almost wrecked her life.

In her last position, under Dr. Bailey as supervisor of drawing in Braintree and Marion in Massachusetts, Cordelia suffered a complete nervous breakdown and spent several months at a sanitorium in Jamaica Plain before she was well enough to return home. At thirty-nine her teaching days were over. And the spring time of her life was ended.

For twenty-five years the schoolroom had been at the center of her existence, first as a student and then as a teacher. She had neither sought after nor desired much beyond what she had found within the halls of learning, and she had always been most happy when matching her own enthusiasm to the eager curiosity of the boys and girls who had come to her. But what could she do now and where should she turn to find a way of life to replace what she had lost? She had driven herself beyond the limits of good judgment, and her reserves were exhausted.

On November 17, 1904, Cordelia Johnson Stanwood came home to the little Cape Cod cottage on Beckwith Hill, a woman still comparatively young at thirty-nine. She was ill but undefeated, confused but not confounded, disillusioned yet not lost. Though she could not know it, there was still a half century of life awaiting her, by far the better part, beyond the spring.

4

Home to Ellsworth

Home is the place where, when
you have to go there
They have to take you in.
Robert Frost

Cordelia's homecoming created consternation and confusion in the Stanwood household. Her sisters had all left and her mother, her father, and her hostile brother Harry were left to cope as best they could with rearrangements of sleeping chambers and family routine, as well as some deeper and more difficult rearrangements of their relationship with a woman who had become strange to them.

Her father, now retired as a sea captain, spent much of his time with his salt-water friends in the boatyards of Water Street, but he felt puzzled and disturbed by this strange cultivated woman who had returned to his fireside. He did not have the solid planks of his own ship under his feet in this storm, and he found it utterly impossible to recognize his first-born child in this nervous, sometimes hysterical, woman who looked well, yet wasn't. In his own hard-headed practical outlook things were either black or white, but this picture was all gray, and fuzzily gray at that. He couldn't make up his own mind whether he should laugh it off, get mad about it, or just keep his mouth shut. He chose to do the last.

Mother Stanwood, however, could not find any such simple answer. Gentle and quiet, she was deeply loyal to all her children, and without any knowledge of nervous disorders she still could understand and sympathize with her daughter in the ordeal she was going through.

With Harry it was a different story entirely. At twenty-five he was too much absorbed in his own ever-expanding activities to allow himself to be distracted by an ailing sister, especially that oldest sister best remembered for her irritating ability when he was growing up to see through and puncture many of his most ambitious schemes for free-wheeling independence. Of course, the only son of the family could always count on his mother anyway. He and his father maintained a semi-polite indifference to each other. Harry was not a sailor. His one voyage on the good captain's ship

View of Ellsworth, Maine, above the dam, in the early decades of the century. (CJS photo)

had terminated about the time they had cleared the channel of the Union River out of Ellsworth. Somehow, they never could get along.

Cordie was not unaware of the disruptive effect of her presence within the family, yet there was little or nothing she could do about it other than to stay by herself in her misery. She was never one who found it easy to confide in others, and except for her mother there was no one here who could imagine how she felt.

In lonely solitude she looked out upon the bleak gray of a November landscape with unseeing eyes and indifferent mind. These were the dreadful days of torturing headaches and nausea, the long hours of depression and mounting tension until the glands in her neck became swollen and inflamed beyond bearing, when her mother would call Dr. King to come out to administer a sedative. Throughout the early weeks of winter there was neither medicine nor kind word which could reach or alleviate the awful burden of loss and failure which bowed her proud spirit.

Many such illnesses have their roots in mental disturbance of one kind or another, and her crippling sickness began to abate only after Cordie's own keen intelligence once more asserted itself and demanded a rational approach to her problem. Just when this began she could not tell, but she did remember that one day while looking down the long hill below the house she had become aware of the rugged peacefulness of what she saw: the flat smoothness of snow-covered fields on either side of Card's Brook, the lazy blue smoke curling up from chimneys in the town, the purple outline of the hills in Dedham silhouetted against the rosy tints of late-afternoon sky. Suddenly she had felt as if a great weight had been lifted from her shoulders, and when a gull sailed majestically through the golden rays of late sunshine she had smiled and whispered to herself, "Oh world, you are there after all. You haven't changed. It is I who have been away, and you have been waiting for me all the time."

Soon after this, simple little things long unnoticed or forgotten as a part of her childhood began to intrude pleasantly upon the grimness of her sickly broodings. Minor in themselves they were little stepping stones back to emotional balance and renewed awareness.

The simmering of the iron teakettle on the kitchen stove, the flickering flames on the ceiling when the lids were lifted for another chunk of birch or maple from the woodbox, the comfortable creak of her mother's favorite rocking chair by the west window in the dining room, the strong, pungent smell of tobacco from the Captain's T D pipe, the rich red and blue against a background of brown in the Aubusson rug on the parlor floor, the opalescent sky blue in the vases on the living room mantle catching and reflecting the light from the hanging prism lamp. All were sensitive responses to homely simplicity, healthy and unencumbered by worldly cares or personal frustrations. From moment to moment to be alive and to be glad about it was all that mattered. This was her most effective medicine.

In the night when she could not sleep she learned to lie quietly and absorb the sounds of darkness--the wind prying with icy fingers around the eaves, the metallic snap of a frost-encrusted nail on the shingled roof, the muffled patter of a mouse in a closet partition. From below in the living room she would hear the subdued creaking of the household heater well stocked for the night with heavy pieces of beech and maple. In the darkness the slender birches were outlined against the spruces and pines like sharp white icicles in the silver-blue winter moonlight.

The slow, cold days of winter piled themselves one upon another like the ever-deepening drifts of snow. She talked seldom, avoided neighborhood gatherings, and hardly ever left the dooryard. She was no longer accountable to persons, places, or things and was answerable to herself alone. Only through what she herself believed and acted upon would she find her way, and that way was waiting in the simple, uncomplicated atmosphere of her home and the woods and fields which surrounded it. It was even reflected in the quiet pleasure of sharing a cup of tea with her mother and knowing that nothing more than her presence was required of her.

Spring seemed to hurry along impulsively in 1905. Southerly winds and warm rains quickly dissolved the high-piled snowbanks into gushing streams which overflowed the west swamp and flooded Stuart's Meadow at the foot of Beckwith Hill. The long, slow winter was over.

Cordie was anxious to pull on high boots and begin the first of what would become countless woodland tramps beyond the pasture bars. Forty-odd acres of Stanwood land to explore. To stand in the vegetable garden and watch the busy energy of the honeybees in the hives at the edge of the orchard gave her a sense of belonging. Close by in the barnyard a few laying hens scratched industriously for scattered kernels of corn while far down the north slope of the open pasture the family cow and a few sheep grazed contentedly. Out beyond the hayfields lay the dark outline of the mixed growth of the forest, the mystery of the swampland, the challenge of the rocky hills. Once beyond the fringe of trees behind the barn, she stepped into another land--a fresh and exciting wonderland which drew her on and on toward new discovery and enlightenment with each bend on every path she followed.

She wrote down her excitement and anticipation:

I love the tangle of moss-green branches that interlace until the woodland looks like a mysterious labyrinth and feel sure if I could but pierce the tangle I would come upon some enchanted castle. I have felt so, and looked for it, ever since I was a child.

Birdsacre, the Stanwood homestead in Ellsworth, Maine, as it looked at the turn of the century. (CJS photo)

From a more practical point of view Thoreau's robust words in *Walden* were appropriate to her position: "I went to the woods because I wished to live deliberately, to front only the essential facts of life, and see if I could not learn what it had to teach, and not, when I came to die, discover that I had not lived." Thoreau's rebellion was her rebellion, equally profound and individualistic, but hers was wholly feminine. She would not, like Thoreau, mount a podium to lecture fellow townspeople, nor go to jail for non-payment of a poll tax over a principle; yet her aggression and dedication, her intense observations and insights, would have a close kinship to his.

Following the pasture fence in search of the shy, elusive thrush, she felt the springy softness of the mosses and lichens under her feet and sought to know them by name. She also watched for the flowering of the fruit-bearing shrubs and spied out the painted trillium, the beds of wild violets, the trailing vines of the arbutus. Overhead the trees spread their branches to shield her even as they created the natural havens for all creatures which lived here. The very stones in the fields and on the hillsides had a message of time and change. Her eyes excluded no vision and she heard the symphony of all sounds, high or low, sharp or muted, within the range of her enlarged and opened senses. This was not a new direction for Cordelia Stanwood. It was a reaffirmation of what she had known in childhood. Now she was free to devote herself wholeheartedly to this world.

She had been only fourteen when she went to Providence to live with Aunt Cordelia; she soon discovered that at thirty-nine she had very little in common with her contemporaries in the Ellsworth community. Most of those she had known as a child were married, had families, and were involved in social activities which did not now, nor ever could have much appeal for her. Many people are naturally gregarious; Cordie Stanwood was just as naturally solitary.

The woodland of Birdsacre. On the back of this print Cordelia Stanwood has written, "Stump on which the Winter Wren bobbed before flying to his nest in the upturned root." (CJS photo)

Knowing this so well herself, she wasted no time in self-pity or bitterness. She turned to her books, her painting, the arts and crafts she had pursued over the years, and to an honest appraisal of what she already knew and a hard look at what she needed to learn to make both an honest and scientific approach to nature. Years later she would say:

> *We bring back from the woods what we carry to it--give little and we get little in return.*
>
> *The friend that you have associated with from your childhood days, the friend that you have found faithful and true at all times, the friend that you have summered and wintered with, is the one that you love best, and when you meet Nature under all circumstances, when you have learned to interpret her varying moods, learned to go to her for peace and consolation; then, and only then, do you know her.*
>
> *Nature will have all of your attention or nothing of you. If you think of yourself, or your affronts, you forget to look for the color, you lose the Cecropia's cocoon, you fail to see the bird or hear the song, and the flowers lose their fragrance.*
>
> *The outer world does not exist to one who steeps himself in thoughts of self. Give undivided attention to that outer world and one's own affairs sink into insignificance.*
>
> *Yes, if you would study the secrets of nature, you must wear stout boots and a strong dress, and you may get a freckle on your nose and become thoroughly tired; yet, from all this comes the thrill of discovery, the knowledge of achievement, and the peace and joy of your own new world.*

Herein lay the fulfillment for that unquenchable curiosity and enthusiasm which had led her to master what others would observe only casually or with indifference.

"Intimacy with nature," Cordie points out, " is acquired slowly. It comes not with one year out of doors or with two."

> *You look and listen, bewail your stupidity, feel that you have acquired little new information; yet, are determined never to despair or give up. All at once you know what you never dreamed you knew before.*

In the beginning, the study of the feathered folk is a delightful torture. There are such a variety of calls and melodies and so many songsters to become familiar with that the novice confounds the call notes and airs of one bird with those of another. If he is content to know just the robin, bluebird, song sparrow and a few others by sight and song, he gets a mild sort of pleasure from his intercourse with the birds, but if he wishes really to lose himself in this world, he must not only work, but work intelligently. When the bird lover has once mastered the vocabulary of the feathered people he begins to be truly in touch with them.

Then as he steps into the woods, it seems as if an invisible curtain drops down behind him and he is in another sphere.

Maine's state bird, the chickadee, became almost a trade mark for Cordelia Stanwood, who photographed these just as they were ready to leave the nest. (CJS photo)

Now that she was seriously committed to the study of birds Cordie would master this vocabulary of her feathered subjects so thoroughly that, like her teacher friend in Greenfield who had identified the black-throated green warbler by his song alone, she too would learn to recognize all the birds of her home territory. Wandering day after day through the aisles of the forest she sought them out. And as she observed and her knowledge broadened she began those field notebooks that would become the source for her many published stories about birds, their habits, their courtship, their nest life.

She selected the standard school notebook for recording her observations. It had lined pages bound in durable brown manila covers, and its seven by eight-and-one-half inch size made it convenient to carry when she went to the woods. That first notebook in 1905 marked the beginning of a pattern of study which would be continued for forty-eight years and lead to two thousand pages of valuable observations.

Cordie the impetuous girl of fourteen on her way to Providence, Cordie the teacher moving restlessly from school to school, Cordie ill at thirty-nine returning home, and now another Cordie, the capable, dedicated interpreter of the great outdoors she had always loved, whose curious mind and tireless energy would unlock secrets others had overlooked or failed to find.

Amusedly she likened herself to her pupils in the Messer Street School of Providence back in 1887. She knew that, first of all, she must master the ABC's of bird language before she could appreciate and understand how every bird is a part of, and fits into, the cosmic whole of nature's wonderful patterns. In fact, she recognized that, either consciously or intuitively, she had always approached nature this way.

"It may well be," Cordie writes, "that people learn to know nature in different ways."

> *If I had not watched for the buds of the arbutus to unfold when a child, if I had not gathered it as a woman, if I had never seen the linnea carpeting the ground, trailing over the rocks, making a flower garden of each old log, and filling the vast woodlands with haunting sweetness, I should care for the flowers less. It is because my mind is stored with groves, fields, swales,*

hills, and mountains where ferns, flowers, and trees grow in unusual beauty in natural surroundings, it is because each bird voice means so much to me and suggests such varied experiences, that I love nature so much. I remember so many cities and towns because of the birds I first saw or heard there, and because of the flowers that I first found growing there. I remember Brookline for its birds and flowers, just as I remember New York for its Gothic cathedral, hand-carved pews, stained glass windows, mural decorations, its theatrical stages, and its Tiffany's.

Cordie's home acres became her schoolroom and her cathedral. She named them Birdsacre.

5

The Spring

. . . its waters, returning
Back to their springs, like the
rain, shall fill them full . . . .
Longfellow

When Captain Stanwood built his house on Beckwith Hill in 1850 to provide a home for his mother, two older sisters and their children, a dug well just beyond a young crab-apple tree in the dooryard supplied water to both household and barn. Although never piped into the house it served the family for many many years. Not long after Margaret Susan, Cordie's mother, came to live in this hillside homestead in 1864 she discovered a "boiling spring" down in the pasture near the base of the hill, a spring that would become a focal point from which most of her daughter's nature studies would later radiate.

The spring itself is remarkable for several reasons. Always over-flowing, its cold water still steams a glass on the hottest summer day and it has never been known to ice over in winter. Located in a mixed growth of pine, spruce, maple and young balsam, it is a favorite haunt for many birds which come here to drink and bathe. Actually this spring neither boils nor bubbles visibly, but its cool sweet water wells constantly from the rocky bottom to feed a little stream which meanders lazily among the ferns and mosses.

Cordie had played around it as a child. This was the spot she had visited on that August birthday in 1879 just before she left for Providence to live with Aunt Cordelia Johnson. When, in later years, she worked so closely with young birds, it was here that she would bring them to fend for themselves and be on their way.

In over one hundred years since Mother Stanwood first found the Boiling Spring it has never ceased to flow, and the birds still come to drink and bathe at its outlet. For half a century Cordie disdained the dooryard well and would accept no drinking water other than that which was drawn from the spring's ledge-cooled source. Summer and winter, in sunshine or rain, through snowdrifts and over ice she came here for her water.

She has even admitted to going there loudly whistling Yankee Doodle when she suspected that gunners might be about in the fall hunting season. But woe to the hunter found upon Birdsacre trails. Her wrath was monumental when she confronted this kind of intrusion, and a lady-like tongue-lashing by Miss Stanwood was an experience no trespasser would soon forget.

With Cordelia Stanwood there was never a "right" time to become involved with nature. For her it was all the time. That curving path down to the Boiling Spring told her story over and over again through the adventures that had come to her along its familiar way. She was there in the springtime, in the heat of summer, in the golden glory of the autumn, even when the snow was deepest and the cold extreme.

Dec. 22, 1906 -- When I went to the spring this morning a partridge was there drinking.

Jan. 19, 1907 -- When I went to the spring today there was a large flock of pine siskins and redpoll linnets feeding on the cones of the larch. When the flock arose from one tree to fly to another, the air resounded with the flicking of their many wings.

Jan. 29, 1909 -- Returning from the spring with a pitcher of water the crisp fragrant air made me glad to live this late afternoon in January. The maples, the elms, and the willows made graceful dark silhouettes against the sky when a flock of snow buntings flew from the street and drifted to the top of an elm. The western sky added the needed touch of color to the white landscape rimmed with evergreens. Long bands of gold extended away from the spot where the sun had so lately been before it slipped behind the hilltop, and gauzy little fluffs of pink and gold floated here and there across the green-blue expanse above. This I treasure as one of the many charming winter sketches that cheer dull places.

Feb. 13, 1910 -- Fluffy, immaculate snow almost to the top of my hip rubber boots. Seeds from the spruces and birches showed on the surface of the snow very plainly. The boiling spring and the brook below looked very

A great horned owl that had been caught in a rabbit trap. (CJS photo)

inviting in the midst of the whiteness. A partridge flew away and chickadees were calling. All the evergreens were bending low with their fluffy burdens, and the hardwood trees were encased in glistening ice. Every breeze filled the air with snow as if we were still indulging in a blizzard.

One feels her presence here more than anywhere else, expressed through those thoughts and feelings that she wrote down over the many years she was drawn to this favored place. Here there was peace and beauty, and here there was sanctuary and comfort where the confusions and frustrations of life could be cast aside and forgotten. She would reach far beyond the spring in her search for knowledge and understanding, but she would never forsake it, and she always returned.

May 20, 1910 -- Saw Haley's Comet this evening. Tonight the robins and thrushes are taking advantage of the moonlight to come to the Boiling Spring, the swale, and the brook to drink and bathe. They seemed to come in pairs and signalled each other all along the way. It was so cold and damp I did not linger to see if other thrushes joined the company.

Dec. 6, 1910 -- Again today the Golden-crowned Kinglets appeared to me when I went to the Boiling Spring. One mite came down to investigate, hovering in air like a green gray ball of feathers with vibrating wings. The orange gem on his head glowed in its setting of jet.

May 13, 1911 -- Went down about 7:30 P.M. and sat at the Boiling Spring. The stars came out and the hermit and veery called in the gloom. The moon was not high enough to light the woods much. The Hermit Thrush sang a few strains, then all was still as the dry leaves uncurled in the evening dew.

Feb. 4, 1913 -- The Arctic Three-toed Woodpecker was at the Boiling Spring, and one lone partridge was trying to uncover the feeding spot.

May 2, 1913 -- Went to the spring with a slice of cake and set out for bird land.

May 17, 1913 -- This morning I was up at 5:00 A.M. Some dried fruit, a bit of cheese, and a slice of bread and butter with a cup of warm milk sufficed for breakfast. A draft from the Boiling Spring as I passed supplied me with all the beverage I needed for the morning. Then off for the woods in earnest with a similar luncheon in a candy box. Old clothes kept me from worrying over my appearance.

June 9, 1917 -- Went to the Boiling Spring this evening about 6:30. The sky was pale blue flecked with fleecy white clouds. As the twilight deepened the birds came down to drink and bathe. The white-throats are the most prominent singers. The thrushes are not singing as much as usual tonight.

Jack-in-the-pulpit near the spring at Birdsacre. (CJS photo)

Feb. 6, 1920 -- Yesterday evening it began to snow and this morning it turned to hail. I should say the snow is two feet deep on a level and so dense it was very difficult to walk. I think it took me all of a half hour to go to the Boiling Spring and back. Then I was so exhausted

I had to lean against the barn. [This storm continued through the night and the next day.] Four teams passed the house in the afternoon, the first . . . in two days. The Postman turned back at Pierce's Corner. Saturday and Friday he turned back at the foot of the hill.

Feb. 11, 1920 -- Everything draped deep in fluffy snow. A wonderful picture at the Boiling Spring.

Oct. 3, 1922 -- It was a wonderfully warm, misty day in the woods, just enough breeze stirring to send down showers of red and yellow leaves. The ground under the pines was carpeted with yellow. Red leaves and yellow needles flew all over me as I walked down to the spring. The withered hay-scented ferns were very fragrant, a few stood erect in the shade in a glory of cream white. They were so exquisitely dainty and beautiful I almost held my breath when I looked at them. It seemed as if a breath of air might transform them into spirits that would float from sight.

Oct. 9, 1923 -- Went to the Boiling Spring this mild, sunny morning. The trees were covered with brown, yellow, and scarlet leaves which drifted down softly to blanket the ground under a rich, crisp carpet. The Ruby-crowned Kinglets were sputtering and singing among the trees as gaily as on a spring morning. Such a delightful surprise.

Perhaps the most revealing among these strongly personal moments of happy communion and response to nature found by the spring occurred on July 2, 1909. As she walked down the trail in the early evening, hermit thrushes, a family of ovenbirds, and some vireos came to drink and feed on mosquitoes and black flies.

There was such a babel of young bird voices I crept near to see what was going on. A pair of ovenbirds were teaching their youngsters to feed themselves. They crept cautiously through the grass, leaves, and twigs with heads down and forward and tails down, chirping softly to hold the attention of the family. Then a parent would fly just ahead of the young with a tempting worm. Such a pretty little game of tag! As I started

Yellow-bellied flycatchers about eleven days old. Cordie captioned this photograph "The Head of His Class." (CJS photo)

for home, the mother crossed the path with a nestling in tow. When I neared the young bird it stopped to catch a mosquito. I knelt down, killed a mosquito, and presented it to the little ovenbird on the tip of my finger. After hesitating for a moment, it snapped it up. So then I put my hand flat on the ground and let the mosquitoes bite while the little bird walked over it repeatedly and snapped up the insects. Finally he answered his mother's chirps and vanished beyond the path. Since I have visited these same ovenbirds many times, I presume they are less timid than those of the deep woods.

Small wonder, indeed, that this woman should find her greatest joy in life where wildness became trust in the hand of friendship.

6

Cordelia Stanwood: Naturalist

I meant to do my work today
But a brown bird sang in the apple tree
And a butterfly flitted across the field
And all the leaves were calling me.
Le Gallienne

Those who have achieved eminence as ornithologists were all very young when they started, and more often than not it was one particular bird that triggered their interest.

Roger Tory Peterson in his *Bird Watcher's Anthology* points out that ninety per cent of the fellows of the American Ornithologists' Union were launched on their careers before they were out of their teens, some when only ten or eleven, and that Audubon, Chapman, Griscom, and Wetmore were even younger. A scarlet tanager sparked a lifelong passion for Elliott Coues; John Burroughs was captivated by a black-throated blue warbler; Julian Huxley fell under the sway of a green woodpecker, while John Kieran succumbed to an ordinary white-breasted nuthatch.

Cordelia Stanwood at thirteen did an oil painting of a frustrated barnyard hen attempting to mother a family of white ducklings. Although neither hen nor ducklings would represent her own "particular" bird, the painting did suggest an interest which would eventually place her somewhere between Audubon and Zim in ornithological history.

One's position in that hierarchy is a matter of degree. Cordelia could not, like some, seek out one individual species for close study and detach herself from the attraction and fascination of every other aspect of nature while observing it. Never was it the bird alone. The caroling bird, the soaring trees, the nodding flowers and swaying ferns, the slanting rays of sunlight on a forest stream all blended and harmonized in a meaningful whole. Her field notebooks and her manuscripts, published and unpublished, constantly reveal this quality. The more one delves into her work, the more apparent this broad dedication. And Cordie arrived at this total commitment step by step through patient, methodical procedures. Never hurried, never deflected or distracted by outside influences, she carried on.

Once she had learned to identify birds by their markings, shapes, sizes, color and voice she joined them in their natural haunts and began to locate their nest homes. Her eyes were keen and her patience inexhaustible. Few indeed could compete with her in outwitting the diminutive winter wren, the jaunty nuthatch, or the darting warbler. At the peak of this her study period she was frequently involved with ten to a dozen bird families at a time from before dawn until darkness brought stillness to the woodland.

The forty-odd acres of the Stanwood property became an open book of vibrant life for Cordelia. She would, of course, range far beyond its boundaries in her quest for knowledge and understanding, but on this soil she was always at home, always sure of what she would find and where to look for it. Every square foot was firmly imprinted in her memory, as sharply defined as the flower patterns in the living room wallpaper of her home.

Open fields, pasture, wood slopes, rocky embankments, swamps, swales and bogs, clear springs and tinkling streams against a background of dark forest encouraged the birds to settle here. Under the hill by the Boiling Spring the thrushes and warblers came to drink and bathe, and flycatchers concealed their nests in the upturned roots of fallen firs in the marshland beyond. Along the north shoulder of the hill the ovenbird sent forth his deceptive call, while far back on the edge of the long swamp the woodpeckers, nuthatches, and winter wrens brought sound and motion into the stillness of the deeper woods.

Cordie became quickly adept in locating favorite nesting sites, and only the meanest weather could interrupt what became for her a daily habit. And it was not a matter of an hour or two now and then. It meant getting up with the sun, walking great distances, sitting quietly for hours, enduring black flies, mosquitoes, and "no-see-ums," often coming

A pair of baby chipping sparrows. (CJS photo)

in wearily after the sun had set. Bug bites and bramble scratches, sunburn and wet feet, twigs in her hair and tears in her skirt, she accepted without complaint. She was doing what she most wanted to do, and she was doing it very well.

Cordelia Stanwood would not necessarily add any revolutionary discovery to ornithological research, but she did provide detailed and exhaustive studies of nesting life that went far beyond what others had done. Within three years between 1905-1908 she had located the active nests of nearly a hundred species of birds, and with infinite patience followed every aspect of their histories from construction and incubation to the first timid attempts of the fledglings in flight. About nest finding she noted:

> *When a man seeks treasure he seems to take it for granted that he must delve in far away fields. Not so in bird study. Plan a long walk, betake yourself to fresh areas, and you may return to find the "gem" you sought within a stone's throw of your door.*

Following the discovery of a chestnut-sided warbler's nest she recorded:

> *When I examine the fragile, gem-like eggs from which these rare creatures emerge and observe the fledglings flitting through the trees in a little over a week from the time they pip the shells, I feel as if the day of miracles is still at hand. It makes every thicket, every hedge, every woodland a wonderful spot since I do not know how many such miracles every briar patch may conceal.*

And later:

> *As I walk abroad in the spring when the birds are coming, as I ramble afield when the birds are nesting, as I saunter through the woodlands when the birds are departing they charm me with their graceful movements, brilliant colorings, their exquisite songs, their fascinating nurseries, and their mysterious migrations. I know that if I spend my lifetime studying them I should discover but a small part of that which is to be learned, and I so feel the joy and the beauty that these tiny creatures scatter broadcast in the world that there is constantly in my mind a deep feeling of gratitude to God for their creation.*

What she saw and heard went directly into her writing, offering vivid testimony to her strong response to every aspect and mood of nature. For example, as she walked home in the evening of June 23, 1911 after a long and busy day among the nest homes of warblers and thrushes she was moved by the beauty of the sunset:

> *This evening the northern sky is full of long, dark, gray-blue clouds. The sunset is one of those gorgeous banded effects that come at this time of the year--yellow, deep orange and purple. A few warblers, the black-throated green and the magnolia, sing in the twilight hush, occasionally a song sparrow, but the wild sweet tranquil songs of the white-throated sparrow and the hermit thrush dominate the evening. The leisurely measures come first from one, then another. One fancies that all is still for the night when the sweet wild whistle of the white-throat or the bell-like peals of the hermit*

> *ring out again, and it is only when it is quite dark and the fireflies begin to twinkle that the last song dies on the evening air.*

So ended a joyous day after hours of arduous involvement with three bird families, the Nashville warbler, the red-eyed vireo, and the black-throated green warbler. Close to each she had built a brush blind so that she might watch more closely the activity around the nest. This was the familiar pattern for most days of every nesting season. On the morning of June 19, 1910 she had set out as usual to record her findings, and with slight effort had identified thirty-one species of birds before lunch time. Even as she did this her senses were constantly sending little flashes of sensory information to her mind: the withered blue-eyed grass, white clover in bloom, strawberries ripening, blackberries blossoming, and the azure glory of wild forget-me-nots painting the border of the brook in the pasture. The fragrance of the linnea was everywhere about her as she recalled the first firefly she had seen the night before.

"The oven-bird," she wrote,

> *was not on the nest this morning when I heard the male partridge drumming in the distance. Three young hermit thrushes were flying around with their mother. I think I could have caught one had I chosen . . . . The nest of the Nashville warbler under a spray of ground juniper containing five eggs was empty . . . . At seven o'clock this morning the eggs in the magnolia nest were still unhatched.*
>
> *Later I found the nest of a chestnut-sided warbler containing young, but the mother bird continued to brood so I did not disturb her. . . . Not far away was a red-eyed vireo incubating three eggs. The nest was rather roughly built as the wet weather seems to have destroyed the willow cotton on which these birds are so dependent. In this same corner of the swamp were the nests of two alder flycatchers. One nest was empty but seemed to be quite new, the other contained four white eggs marked with a few light and dark brown spots on the larger end.*

Seven families at various stages of development in one day!

Most observers would have been thrilled with one or two.

Some of Cordelia's excursions to the woodland, however, were far less pleasant than this June day. For example, on May 14, 1911 when she took a field trip to find the yellow palm warbler:

> *Today I walked nearly three miles to the swamp where I found the yellow palm warbler last spring. I heard a number sing but saw only one. I never saw such a dreadful place to visit. It was covered with quaking bogs, overgrown with birdwheat moss and sphagnum, and cut up with muck holes covered with liverwort. Trees had burned and fallen, many had been uprooted, and the land had been partly cut over in frozen weather so that the entire area was submerged in brush.*
>
> *I feared at any moment that I might sink from sight as I plunged from one insecure foothold to another. Talk of the swamps of Florida! This morning I wondered if they could be more inaccessible. After a very long time this led me to a swamp grown over with white birch, alder, witherod, black alder, and a few larches. This growth was not dense and the dead swamp grass was knee deep. The railroad ran along one side. Here I found the yellow palm warbler, the redstart, the chestnut-sided warbler, and the northern yellow-throat. All the meadow is still very wet. A horse would be easily mired.*
>
> *A large brown bird rose from the marsh beside the tracks. It was a bittern, but it made no noise, rose only a short distance, and lumbered just above the low trees until it vanished from sight.*

Although Cordie was content to spend whole days with only the wild creatures for companions, she was by no means a hermit. Many who knew her have pointed out her charming way of infecting others with her own dedication and enthusiasm whenever they joined her in the woodland. In her writing the facts are recorded with meticulous care, but always couched in a language of feeling and emotion that illuminates the incident. The dimension of the sensitive human observer is always present. She might well sit for long hours beside a thrush's nest and record every detail of the activity around it, but her ear was ever attuned to the larger harmony far above the earth-bound structure on the ground. At the close of such a day she takes one with her to share her own joy and thankfulness to be alive in such a world:

> *Stand in the dim aisles of the forest in the twilight when the sun shows orange and crimson through dim vistas of interlacing branches and listen to the hermit thrushes. They perch at different heights on the side of the woods illuminated by the setting sun, and vie with each other in hymning the glories of the universe. Each peal of melody is more indescribably perfect. Before the last cadence of one song dies on the air, a pure, serene,*

The hermit thrush or "swamp robin" was always one of Cordie's favorite birds. The thrush in the photograph is about 20 days old and is a tame bird that was returned to the woods. (CJS photo)

> *exalted paean of praise ascends from another golden throat. The air palpitates with thrush refrain. The song is spiritual, tranquil, and unspeakably beautiful; it is the embodiment of repose; it has the power to evoke man's better self; it calms the turbulent spirit with its "Peace, be still."*
>
> *To know and love the hermit thrush--the Voice of the Northern Woods--and to receive his benediction in the twilight is one of the privileges Nature confers on those who worship at her shrine.*

This gifted woman's deep and lasting conviction of personal identity with all things in nature is a vein of gold running through the pages of objective observation. It is that same rare quality so often found in the writings of Burroughs, Muir, Seton, and Thoreau. Cordie's emotional, even religious responsiveness to Nature by no means clouded her eye for accurate scientific observation. Compare her idyllic paean to the hermit thrush, just quoted, with her account of the same bird in an article printed in *Bird-Lore* in 1910:

> *The nests are very much alike. The outside of the structure is composed of moss, dead wood, twigs and*

The nest of a hermit thrush in a patch of dwarf cornel (bunch-berry). (CJS photo)

> *hay; it is lined with a small amount of black, hair-like fiber, and pine needles. Once or twice the foundation of the nest consisted of more than the ordinary amount of moss. At another time it was made almost entirely of sticks and twigs. Fourteen were lined with pine needles, one with red fruit stems of bird-wheat moss. The proportions of all nests are about the same. The one constructed entirely of twigs was about a half-inch thicker at the top than the others . . . . The Hermit Thrush usually nests in open spaces in an unfrequented wood,*

beside a woodroad or even a quiet street, and on the borders of pastures skirted by woodlands. The nest is placed, generally, under a low fir tree, occasionally under the tip of a long fir branch, rarely in a clump of ferns. Seven nests were located in a knoll, two in a damp hollow, and six just above the swale in the dry earth of a hillside. In almost every case, the slight excavation for the foundation of the nest was made in the loam of a decayed log or stump.

And again, the range of her detailed observations and scientific accuracy is vividly illustrated in a paper she did for *The Auk*, the official publication of the American Ornithologists' Union, in October of 1910:

A SERIES OF NESTS OF THE MAGNOLIA WARBLER

The warblers were late in 1907. The cold, backward spring was behind time in unfolding catkin and leaf whereon the insect hosts prey, and the warblers who live on the insect life keep pace with the resurrection and birth of moth and butterfly, mosquito and aphis, caterpillar and beetle. It was the 17th of May before I heard the weechy, wee-chy, wee-chy; *or the* wee-o, wee-o, wee-chy; *or the* wee-chy, wee-chy, wee-chy-tee *of the Magnolia Warbler, and all of a week later before I saw one. After that they came in flocks, those gorgeous, floating flowers from their winter homes in Panama and Mexico.*

The Magnolia is one of the most beautiful of the birds that comes to nest in the cool north. While migrating the bird is noticeably restless, even for a warbler, keeping well hidden within the evergreens where it feeds much of the time, although it makes frequent excursions to the larches, gray birches and other trees of the swamp and its surrounding woodlands.

On the 13th day of June, I took my luncheon for a day in the woods, not that I was going far, but the days are all too short when birds are migrating and nesting, and I was bent on hunting birds' nests. Towards noon my efforts were rewarded by finding the nest of a Magnolia Warbler nearly completed. Two days later, I came upon a second nest of the same bird, and six days later a third. On the 15th day of July, I just missed placing a fourth. By accident, I discovered the empty nest later.

All these nests were composed of similar materials,--hay, stems of cinquefoil, a plant fibre resembling hair, horsehair, plant down and spider's silk, yet each one had a character of its own, due to the greater proportion of one or other of the materials used in the nest, and the way in which the nest was placed in the tree.

The first nest was the most exquisite Magnolia Warbler's nest I have ever found, and I have been so fortunate as to locate at least twenty-five of them. In this nest some hay and the fine tops of cinquefoil served as a foundation, but the greater part of the small mansion consisted of a fine black vegetable fibre resembling horse-hair. So much of this black, hair-like material was used, that when the edge was covered with down from the willow-pod, a person looking at the dainty abode in its setting of fir twigs could see nothing but the jet-black lining and the fluffy, silvery plant-down around the throat of the nest. The structure was partly pensile, being bound with spider's silk to the two branches at right angles to the main stem. The front part of the bottom was supported by the branches beneath. The interior was modelled by the dainty curves of the mother bird's breast. It was built in a small fir two feet from the ground, surrounded by a growth of fir and gray birches.

The second nest consisted mostly of cinquefoil stems, with a few strands of hay, a lining of horsehair, and a few dots of plant down fastened over the exterior of the nest with almost microscopic meshes of spiders' silk. The cinquefoil stems make a very attractive nest. It is so brittle, it cracks every two or three inches, giving the nest a light, angular appearance which is very different from the effect produced by using hay. The dots of plant down, with the almost imperceptible silk veiling, add also to the effect of lightness, yet a Magnolia Warbler's nest is a very substantial little affair. It was placed close against the stem of a fir where the ascending branches form a partial crotch, and was about three feet from the ground.

The third and fourth habitations had the appearance of being shallower. They were made of about equal parts of hay and cinquefoil, and lined with black hair-like plant fibre and a few horsehairs. The outside was strengthened with plant down and spider's silk, and it was safely anchored to the surrounding twigs with spider's silk. One of these nests was placed on a forked branch near the end of a long spruce bough some three feet above the ground; the other between the extreme tips of the branches of two little fir trees, at about the same height as the former.

An unusual nest of the magnolia warbler, suspended between two fir seedlings. (CJS photo)

A typical nest was about 1¾ inches wide inside at the top, and 1¼ inches deep, the bottom a half inch thick, and the walls at the top three fourths of an inch thick. All the nests somewhat resemble in shape the bowl of a spoon. In three nests there were four cream-white eggs in each, with the pinkish tinge that nearly all freshly laid eggs have, spotted in a ring around the larger end with reddish brown, umber, and black. There were minute specks over the entire egg.

The nest of the magnolia warbler, with five eggs. (CJS photo)

In the first nest, which was unique in many respects, the eggs were marked with burnt umber all over the larger end, as if a person had scrawled over them with a Japanese brush.

The eggs were laid on four successive days before 8:30 A.M. On the fourth day the female took up the task of incubation before 10:30 A.M.

If one comes cautiously to the nest while the bird is incubating, the startled little mother usually slides silently into the undergrowth and remains there. Once when I waited by the nest a long time, the bird returned to scold, but kept carefully out of sight and chirped very little. Another bird when flushed from the nest flew to a near-by tree and fell like a dead weight from the limb with (apparently) a broken wing.

In twelve days the eggs "had wings, and beak, and breast."

Five healthy nestlings. (CJS photo)

On the fourth day one of the nestlings opened its eyes, tiny slits, but it closed them quickly as if afraid of the light. The fledglings were burnt orange in color, covered with long, dark brown down. The quills and feather tracts were well indicated.

Feeding time for the baby magnolias. (CJS photo)

On the eighth day the nest was empty, but I saw the young birds fluttering through the trees with the parent birds, only a few yards from the nest. Probably the violent rain and thunder storm of the day before had hastened their departure. (The other nests were either destroyed or the young eluded my vigilance.) When the young birds were in the trees near the nest, the old bird exposed herself most needlessly. All her caution seemed to have vanished. It was an effort to attract attention to herself from the young birds, who were immature and noisy.

Although the nests of the Magnolia Warbler were so similar, I had actually to see the bird sitting on three distinct types of nest before I could believe that all the structures were made by the same species. The third and fourth types were sufficiently similar to be identified.

A pair of young magnolias just leaving the nest. (CJS photo)

In 1908, I had the opportunity to make a careful study of four more Magnolia homes. May the 13th, the birds had just begun to place the lining in a nest about two feet up in a low spruce. Both birds brought cinquefoil and black plant fibre to the nest, and entered it to put the materials in place. The female seemed to do most of the work. She pressed the material into place with her breast, moving around gradually, so as to make the sides uniform. When the birds detected my presence, which was almost immediately, they always ceased coming to the nest for a time. The rainy weather seriously interfered with work on this nest. The last material was added six days after the nest was started.

Three days after the nest was completed, on June 5, the bird laid one egg about half as large as the ordinary Magnolia egg. That would indicate that she was a young bird and this her first nest. On the evening of the twelfth day of incubation, an excessively hot day, there were two young birds in the nest. Probably one young bird died from the hot sun rays pouring down upon it while the parent bird was procuring food; the small egg remained unhatched. Of the two nestlings, one was much stronger and larger than the other.

On the third day, the eyes of the nestlings were beginning to open, and the feather tracts were indicated by dark brownish blue spaces. On the fifth day the wing quills were three fourths of an inch long, and the body well covered with pin feathers.

On the seventh day the wings of the young Magnolias were a mixture of yellow-green, black, and blue-gray, with buffy wing-bars. The head and back were dark brown, the breast heavily striped with grayish brown, and the belly was yellowish. On the morning of the tenth day, June 30, the nest was empty. I visited this nest every day for thirty-one days. If my frequent visits did not hasten the exit of the young birds from the nest, it would be strange.

June 3, 1908, I came upon two Magnolias just starting a nest in a fir three feet from the ground. First bits of spider's silk were laid in the shape of the nest on the brush-like needles of the fir. The bird seemed to secure the spider's floss by rubbing it against the twigs with her breast. Later bits of hay or cinquefoil stems were bent in the shape of a loop or swing and secured by the silk. The next step was to bend the material in the shape of a circle around the top, always pressing it into shape with the breast and securing it at intervals with knots of spider's silk. A frame similar to this seems to be constructed by the Magnolias always before filling in the foundation. The birds were three days placing the foundation of hay and cinquefoil, and three days lining the nest with horsehair. I have seen nests that I thought might have been constructed more quickly, so little material was used either for foundation or lining.

The other two nests were similar to those I have described except that one was five feet up, and some of the red, hair-like fruit stems of bird-wheat moss were used in the lining. This was placed between the tips of the branches of two low trees. The bird that built the high nest with the colored stems in its lining, laid the smallest eggs I have ever seen in a clutch of this species, and was extremely gentle. Unfortunately crows or squirrels carried off the eggs so that at this point my observations ceased. The small eggs would indicate they were laid by a young bird, and the somewhat exposed site suggests that she was inexperienced.

In 1909, I found five nests similar to the others, with these slight differences: One was placed seven feet up in the tips of a long spruce branch and lined with coarse dark brown roots such as the Hair-bird uses for the exterior of her nest; another had a middle lining of the fine tips of meadowsweet twigs, which was coarse material for the Magnolia to handle. This latter was placed in the axis of a fir branch two feet from the ground.

"Faithful unto Death." A mother magnolia warbler who died shielding her nest from a severe rainstorm. (CJS photo)

The eggs of this year were much blotched with reddish brown or umber, sometimes in the wreath around the larger end the blotches being confluent; at other times the blotches pretty well covered the larger end or extended far down the sides of the egg.

On the second day of July, 1909, I came upon parent birds with young. Both old birds flew around me, chirping with consternation when I paused to chat with the dainty mite that confronted me on a low fir. The mother spread both wings helplessly and fell from branch to branch and from low trees and stumps to the ground. The male bird contented himself with flying around with his mate and chirping. This would indicate that the male assists his mate in the care of the young after they leave the nest.

Her studies of the golden-crowned kinglet reveal this same scientific observation combined with keen sensitivity:

> *Up among the tops of the large evergreens on a spring morning one hears the golden-crowned kinglets. A little larger than a hummingbird and a trifle smaller than the winter wren they are so tiny and so nearly the color of the foliage that their thin, wiry "zee-zee-zee" call attracts the ear of the casual observer scarcely more than the piping of insects.*

Nine little golden-crowned kinglets. (CJS photo)

> *Most of the nests of the kinglet that I have come upon have been fashioned in the black spruce from fifteen to twenty-five feet above the ground, but these birds normally live and undoubtedly breed in the ancient, mossy forests where white pines, spruces, firs, and hemlocks seem to touch the sky. However, these castles in the air are so cunningly hidden I find it almost futile to attempt to locate them.*
>
> *The nest itself, a beautiful compact structure of mosses, twigs, and spider's silk, lined with rabbit or squirrel hair, and rimmed with feathers, is usually concealed in the thick foliage near the tip of a heavy limb.*
>
> *When the pensile hammock will bear her weight the mother bird hops down upon it and shapes it with her feet, always turning from left to right. Then when she is able to sit in the basket she carefully moulds it with her breast, still turning from left to right.*

The completed walls of this cocoon are nearly an inch thick, and the masses of moss, hair, and feathers woven into its sides and bottom make it a snug, secure incubator for the six to ten kinglets which will grow up here.

The "pensile hammock" of the golden-crowned kinglet. (CJS photos)

> *If one wishes to study the nesting habits of the golden-crowned kinglet he should seek the woods by the first of April, although the snowdrifts still linger and snowstorms may yet be looked for, since this bird appears to build as frequently in April as in May.*

The facts march across the pages of her writing in neat and continuous precision, but Cordie herself is also there, speaking her mind, revealing her heart, searching out answers to life and its meaning. Slowly, deliberately, and with solitary patience she projects herself into the shifting pageantry of color, form, motion, and sound which portrays the mystery of the physical universe. All things had their place, all things were interdependent, growth was constant--balance, power, symmetry, and beauty everywhere. For Cordie it created immortality out of mortality.

> *A mist of smoke, a mist of fog, a mist of willow and alder catkins, a mist of twigs and softly bursting buds, a mist of shadbush and hobble bush blossoms, and a mist of pine, fir, and spruce branches -- that is today!*

Thus it went with her when she "took to the woods" as a way of life and lived in them far more hours than she spent under a roof. In truth, her most natural home had the open sky for a roof with rafters of towering pines and spruces above a floor of brown needles and mosses. The broad fields were sunlit entryways to the shadowy chambers of that "enchanted castle" which had eluded her since childhood. The reader can be startled by a sudden flash of inspiration or poetic insight in the midst of objective factual material, revealing with unexpected clarity the character and personality of Miss Stanwood, the woman, beside C. J. Stanwood, the ornithologist.

For example, on May 18, 1911, this spinster of forty-five had recorded in her field notes that at 8:00 A.M. the temperature was 53° and the wind south-westerly as she proceeded on her way to locate over thirty species of birds, either visually or by song. She had found a few late arbutus blossoms, had whistled at a large, tawny rabbit which sat and looked at her before scampering off into the brush. "The ground," she wrote, "is blue with violets, white with violets and houstonia, and yellow with dandelions. It suggests one of the flower besprinkled foregrounds of Botticelli."

Then her thoughts ranged far beyond this beautiful May day of bird migration to link this present hour with a long-forgotten youthful memory:

> *The woods have the odor of strawberry time. I cannot describe it. When I was a child and went out in the dewy grass and found the luscious, fragrant, ripe strawberries nestling among the hay-scented ferns, I did not notice the odors of the damp earth, the grass, the ferns, and the berries; but, when I fill my lungs with this air--suggestive of the pine, the balsam, the spruce, the white violet and the blossoming willows--I see again the pasture where I knelt as a child to pick berries, where I found so long ago the same kind of hermit thrush nest that I find in these days. And the song of that hermit thrush that I heard as a child and called the Swamp Robin is still to me the most beautiful music in the world, and the nest of blue-green eggs still gives me a thrill whenever I find it. My childhood associations with this bird will always make it seem different from any other--even Cock Robin who covered the Babes in the Woods with leaves.*

Cordie knew and loved all the birds from the dusky crow to the shy winter wren whose song "is high, liquid, wild, [and] suggests the creaking of the naked trees in the high cold blast. It echoes the tinkling and the rippling of little brooks made by the melting ice and snow." But none of them, however captivating, could ever bring to her that feeling of divine peace evoked by the liquid melody of a thrush's refrain at twilight.

> *When the thrush speaks to me, it seems as if the rags and tatters that enshroud my soul fall away and leave it naked. Then I must be simple and true or I cannot feel the message the small voice brings to me. When the thrush sings, I desire to live in a small, scrupulously neat camp, open to the sun and the wind and the voices of the birds. I would like to spend eternity thus, listening to the song of the thrush.*

Cordelia Stanwood always stood in reverence and awe before every expression of life. She read a spiritual significance into the natural life around her. To her the painted trillium was never just another flower, for it "carries at its heart in purple stain the symbol of the Trinity." The hobble bush in bloom became "the candelabra of the Lord, the thank offering of Mother Earth," and the delicate beauty of a white birch in new leaf told her "that Mother Urania had been robbing Father Neptune and appropriated his sea shells with which to deck her limbs." When she stood in Tinker's Woods on a summer evening under the towering pines and spruces that soared eighty to one hundred feet into the sky she would be moved to say that "slender-branched toward the top and far enough apart to give considerable light they form the long dim aisles of God's church in the woodland."

Wherever her footsteps led, "the most common and inconspicuous spray that raises its head by the footpath is arrayed in silken sheen that a princess could not buy" and "the unfolding buds of spring are the ethereal promise of the rich, gorgeous autumn fulfillment."

A bird flashes by to vanish within the shubbery beyond, and she says to herself with a smile, "Isn't it strange that the bird that flies away before I catch a good look at it seems more beautiful than any I come to know?"

Cordie never lost her joy in the perennial renewal of life's rhythms. Fully aware of her own mortality she once wrote these words at the close of a fall day:

> *The leaves fall like feathers from the birches, but once in a while one falls prone and heavily like a grasshopper. Even as I sat there they began to cover me and I thought how little time it would take to blot the remembrance of me out of the world.*

But this blotting out was less a reflection of regret or sadness than a recognition of the continuing life that would spring anew from beneath this blanket of fallen leaves. Actually, her thoughts ran ahead to the season to come when she could say once again, "It is a wonderful thing on a glorious spring morning to be in the midst of a bird migration."

This refreshing optimism and positive attitude toward life culminated thus at the age of eighty-six:

> *Between the two flower beds at the side door this morning I saw a mite of a bird much smaller than my*

thumb. For a time he seemed quite oblivious of my presence as he performed a rapt dance, swinging up and down in a semicircle from a few feet above the ground almost to the top of the roof. I have seen the humming-bird in the spring perform his aerial dance to excite the admiration of his mate, but this time I beheld no feminine audience. The little pagan seemed to perform his ecstatic dance as an act of worship before a beautiful yellow rose festooned with blossoms. Could it be that the hummingbird and Elizabeth Barrett Browning both worship God in the blossoming shrub: "Earth's crammed with heaven and every common bush afire with God; but only he who sees takes off his shoes." Yes, verily I do believe that I caught the Ruby-throat saying his prayers this morning.

7

The Field Notebooks

Go forth under the open sky, and list
To Nature's teachings . . .
Bryant

Better than half of Cordie's adult life is well catalogued in her field notebooks, started in 1905 and continued without interruption until 1953. In scanning these notebooks, one is impressed by the tireless energy of the woman who compiled such a wealth of significant detail. She was so committed to intense observation and so sensitive to the rhythms of natural life that the reader inevitably seems to be sharing the experience with her.

The most intensive and fruitful period of her field studies took place within a comparatively short time, between 1907-1920, the period also of her most productive literary work. During these years she was moreover actively involved in photography, which became a major tool of her research. Her field work, her photography, and her writing were thus synchronized into a pattern that became a positive expression of her life's purpose. The years of formal education and teaching had been preparation for this larger expression of herself in her search for the beauty and truth she now found so abundantly.

For one who wrote so regularly throughout her life and kept such meticulous records of everything she did, it seems rather odd that Cordelia never maintained a personal diary. Notebooks survive on courses she had taken, notebooks on teaching, notebooks on art and literature, a genealogical record of the Stanwoods, Leonards, and Bowns, a scrapbook of correspondence from editors and authors, even an abbreviated autobiographical sketch of herself with emphasis on the teaching years, and, most importantly, the field notebooks. But no diary.

Had there been one, much that now remains puzzling about Cordie Stanwood might be better understood. One is forced to turn to her other writing, and particularly to her field journals, to learn what one can of her feelings, her innermost thoughts, and those ideals and convictions which both directed her actions and shaped her character. Fortunately these notebooks offer cameo pictures of Miss Stanwood which compensate for the absence of a diary.

With a notebook on her knee while alone in the woodland she was closer to the real Cordie than anywhere else. In what she wrote at such times one can sense the inner harmony and joy she knew when among her birds. Some of it is humorous, some serious, and all of it reflects thought and sensitive concern for the world of nature she loved.

When she says that "the purple finch is a harlequin in song" she captures what most feel about this colorful bird but few find the words to express.

On the afternoon of July 1, 1907, she had been roaming through the woods and fields looking for nests. "Today," she wrote, "I had been searching on my hands and knees in the locality where I had heard the northern yellow-throat singing." Then, with tolerant understanding she went on to express her frustration in defeat:

> *The northern yellow-throat still sings early and late, in rain or shine. "Won't you see this? Won't you see this? Won't you see this?" Look high or low, never a nest will she show you. I consider her a most deceptive little bird.*

Admitting failure she had started to go home and was walking across a small knoll covered with alders when she was rewarded by an unexpected discovery.

> *I heard something like the pulsing of the wind that didn't seem in rhythm with the light breeze. I went back and waited. After a time I moved in the same direction again. Again the sound was repeated. I decided*

that it came from a clump of alders with fern and hard-hack growing in front.

She carefully examined the ground and the trees nearby but could discover nothing. "Finally," she says, "I looked through the trees from the open side and just ahead of me was the vigorously constructed nest of the pretty-dear."

Then turning ahead a few pages in this particular notebook one finds the entries which suggest the happy ending:

July 22--The nest of the pretty-dear or alder fly-catcher is full of young birds. The markings on the wings are quite plain.

July 26--The young birds are standing on the side of the nest!

July 27--The nest today is empty. It is fastened to the branch with spider's silk and made of very coarse grass, lined with fine yellow hay.

Turning back to the entry of July 1 one finds that the afternoon had one more surprise for her before the sun would set. She discovered another nest, again by accident and unexpectedly, the nest of the Nashville warbler.

Two white-throated sparrows always scolded me whenever I neared a certain bit of undergrowth. I decided to look for the nest. [Later she would find it.] I went into the midst of the growth and concealed myself to watch for the birds' return. I heard a sound at my right, then to my left. A magnolia warbler scampered past into the evergreens out of sight. A white-throat took up his stand and kept up an incessant scolding. I looked on the ground around me to be sure I wasn't on the nest but saw nothing. Suddenly I noticed a peculiar spot on the moss. It proved to be the wing of a young warbler. I was so excited and so afraid I might have hurt the young bird that I hastily left the young to the care of their parents and went home.

The next morning she had returned early and found "five plump, unharmed, little warblers" sitting snugly in their nest.

"While there is great pleasure in travelling the pathless woods, it involves much physical weariness." In these words Cordie made it clear that she was not indifferent to the hardships she encountered. These hardships, however, only stiffened her resolve to carry on. "Another wearisome day with apparently no results." Then she consoled herself by picking a bouquet of arbutus and went on to say, "I do not mean that I do not enjoy the bird work. I do love it thoroughly, but the results are small, often, compared with the exertion." This she wrote in May of 1914 when she was forty-nine, but small results or not for another forty years she would pursue her observations and build upon her knowledge.

Today we continue to share the perfection of these closely observed moments, however small. Her power to communicate these moments so freshly results from her habit of living with all her senses open and alive. In the quietness of an evening she had gathered a basket of trilliums and was coming home in the darkness across the pasture. "My dress," she explains, "brushed against the sweet ferns and a spicy, pungent odor filled the air."

She labored for accuracy and delighted to correct any misconception or hasty judgment. Surprised one day to observe a hairy woodpecker on the ground apparently eating snow she wrote:

> *I went later and examined the spot and found that a large piece of suet had fallen there. This shows how often that "appearances are deceitful" in the world, and how careful the scientific writer should be to verify all he surmises.*

Occasionally a spirit of comic playfulness dominates over the scientific reporter:

> *A little junco appeared with just two tail feathers, and these so much awry that it looked as if there were but one. I wondered if she had encountered the shrike, and if her chances of matrimony were endangered for the spring mating match. Fortunate lady that she doesn't own a mirror. Let us hope that she doesn't have an intimate and frank friend.*

And she says of a chickadee who came to the window for suet: "As I returned to the room it flew away with a call that sounded exactly like 'Cheese it! Cheese it!' Such familiarity with the slang of the street Arab."

This free translation of bird language appears in her report of an encounter with a family of robins on June 22, 1909:

> *This morning I found a young robin in a tree and picked it up. It opened its beak and let out one loud howl from its golden throat. In an instant the parent birds were upon me, calling "Thief! Thief! Thief!!!" or perhaps something worse. I put the bird down at once, for, I meant no harm, but they pursued me across the pasture with such vehemence I truly blushed with shame.*
>
> *None of the robins in this pasture had ever been afraid of me before, but I just knew that these indignant parents would give me a bad reputation with all birdland this morning.*

Cordie seldom makes any observations about the social, economic, or political activity which went on around her. Clearly she felt little reason to do so, and field notes on birds would hardly be the place to look for such comment, but she was not oblivious to what went on "out there," as we see by this entry on May 18, 1941:

> *Olive-backed thrushes singing a few bars tonight. I am thankful that I have heard the thrush sing once more. The sweet ethereal strains suggest music from a heavenly shore. It seems too sweet for earthly music, and yet it will be common for weeks to come. The world, however, will be so busy discussing Hitler and Hess that it will not hear the glorious singer at all.*

About the time Cordelia was beginning her own nature studies, Gene Stratton Porter heard from a publisher that she should "cut the nature stuff" if she expected her novel *Freckles* to sell. Twenty years later, her story of the Limberlost Swamp in northwestern Indiana had sold two million copies. Today the reading public is far more aware of the importance of "nature studies" than it was when *Freckles* or *The Girl of the Limberlost* came on the market. One can no more "cut the nature stuff" from Cordelia Stanwood's writ-

An olive-backed (Swainson's) thrush bristling while shielding her young from the heat. One of the photographs Cordie took from her canvas blind, using a sheet as a back drop. (CJS photo)

ings than from Gene Stratton Porter's novels. "Everything in nature," says Cordie, "is beautiful taken in its supreme moment." One May day she records such a moment:

> *The combined fragrance of all green things is indescribably delicate, fresh, and new. Just this proportion of alder, wild pear, white violet, and gray birch was probably never before so mixed with dew and given to the breeze . . . on a spring morning.*

Her world is a world in miniature, delicately balanced in the interdependence of all living things, and properly independent of the rush and turmoil which beset the ways of man. She could, however, become painfully agitated and exasperated by the intrusions imposed upon the natural world by man, whether through ignorance or callous indif-

Baby olive-backed thrush.
Two young olive-backs at nest-leaving time.
(CJS photos)

ference. Not the cat responding to its natural instincts, and perhaps neglected at home, but the thoughtless neighbor is the real target of this indignant entry on a February day when she was eighty years old:

> *I do not wish to use vulgar language but it can not be vulgar to quote the poet Milton who said that Satan could make a Hell of Heaven. Surely my neighbor's cat can make a Hell of Birdsacre. Go to town and fetch home suet for the chickadees that flutter against the window begging for food. Look out a moment later only to discover that THAT cat has been up the bush and torn off all the suet. She has also eaten all the cracked wheat bread. She perches on the windowsill or steps a half dozen times a day. She seems to have perpetual hunger.*

The field journals are a constant treasury of the unexpected, and one cannot afford to overlook a single passage for fear of missing something significant.

> *As long ago as the days of King Solomon people were interested in the migrations of the birds for that wise man wrote: "For lo, the winter is past; the rain is over and gone; The flowers appear on the earth; the time of the singing of birds is come, and the voice of the turtle [dove] is heard in our land."*

Dr. Dallas Lore Sharp paraphrased this last line to describe the coming of the golden-winged woodpecker or flicker: "the voice of the flicker is heard in the land." Cordie commented:

> *This spring after a long, cold winter, the flicker appeared in Maine as early as April 2. His strident "Wick-a, wick-a, wick-a" came from every treetop. Finally a pair of these jolly birds selected a dead balm-of-gilead tree in*

Two young flickers, called "High-holes" in Cordie's note. (CJS photo)

the dooryard for a nesting site and before I was aware had excavated a chamber and taken possession of it. So this season the words of King Solomon and Dr. Sharp have come to me with new meaning. My soul has been overflowing with sympathy for the bird lovers of all time who have looked forward to the coming of the birds, God's messengers of new hope to each of us.

Three score years have failed to dim the vividness of descriptive passages that preserve for us a moment of her life:

Robins here now. They sing and call everywhere. The meadow is full of them. Soon after I went out the great orange cartwheel of the sun shone through the tops of the gray birches. The ground was slightly frozen, the water just skimmed with ice, a white frost covered the sere fields, and the city slumbered in the thick white mist from which the steeples and roofs just emerged. All the birds sang. Soon the mist enveloped everything and veiled the bright face of the sun completely. The birds nearly ceased singing, and drops of water dripped from everything.

A six mile walk at eighty-three to place flowers on the family lot in the McKenzie Cemetery gave her reason enough to be out-of-doors under the open sky with her birds. She never needed excuses for what she did, only more opportunity to indulge in whatever brought her greatest pleasure. So much to see, so much to do, so much to know. "Nature," she observes,

does not venture all her fortunes in one cargo. The birds of a species come in waves. They are not all subject to one cold spell or one severe storm. The arbutus does not all open in one day. Some grows on a hillside, some in a swamp, some in the sun, some in the shade. It blossoms during a full month or more.

At eighty-four she paid tribute to another ornithologist who had described the clarion call of the pileated woodpecker:

Heard the Cock of the Woods give his rounded musical whistle. Mrs. Florence Merriam Bailey [no relation-

ship to Dr. Bailey] calls it his "bugle call." I wanted some word to describe it, but all I could think of was a full, rounded whistle. The first time I heard the bird give the call, it impressed me as remarkably beautiful and entirely appropriate. I only wish I had been the one who thought to call that whistle the "bugle call" of the Cock of the Woods.

In less than a dozen words on August 1, 1941 Cordelia Stanwood sharply defined a woman who knew her own measure: "Today is my birthday! I have cleaned house faithfully all day." Through housecleaning she was grudgingly admitting to herself that even at seventy-six she could still make concessions which opposed the priorities in her life. Housekeeping, for Cordelia, was always secondary. With her there was a constant desire to be in the wide open spaces and away from the confinement of four walls.

Then on a sunny day in May in 1953, she had this to say:

I went through the kitchen. The door was open because I had been sitting on my camp stool in the sun in the shed this afternoon, picking over feathers for a pillow. As I returned from an errand in the other part of the house, I heard a peculiar loud buzzing. Lifting a curtain I saw a hummingbird trying to get out through a pane of glass. I got a chair and put it in place slowly, and so slowly, raised the curtain. I put my hand over the hummer and he offered no resistance as I got down. One bright eye regarded me from behind my thumb. Holding my hand against my face I hastened to the japonica with its gorgeous blooms, opened my hand, kissed the exquisite creature on the top of his head and expected him to fly into the japonica. He had each little leg curled up in his soft, white feathers, but made no effort to stand up. Suddenly, the wings began to buzz. Then the dainty mite soared one hundred feet into the air and disappeared into space.

She was eighty-seven years old, and there would be but one more entry, on June 3, in this, the last of those journals which she had kept so faithfully since 1905.

A tame ruby-throated hummingbird named "Baby," on the hand of Eleanor Alexander Shea. Cordie wrote a story about Eleanor and her hummingbird for The House Beautiful *for March 1920. (CJS photo)*

8

Naturalist as Writer

Living Nature, not dull Art
Shall plan my ways and rule my heart.
Cardinal Newman

After I gave up teaching there was a time when I congratulated myself that I was coining money in my spare moments by scribbling, although the majority of my checks were very small. In those days I kept the idea of writing constantly in mind.

In these words Cordelia Stanwood had introduced a review of her long struggle to achieve effective authorship. Good writing had not come spontaneously, and there was a long period of trial and error before she eventually developed a method and style which best suited her personality and led to publication.

From the beginning she kept a voluminous scrapbook of the correspondence she had with other bird people, together with the acceptances and rejections of her articles. One day "out of curiosity I looked over the scrapbook in which I hoard letters of acceptance from Editors, and I was surprised to see that my receipts from writing totalled $1600." This same scrapbook also reveals how much of her material found its way into the works of other writers.

Cordie, however, did not have to find her way alone in learning her craft as a writer. She received encouragement and advice from several sources, some of whom were established ornithologists with literary recognition.

Dallas Lore Sharp, Boston University professor, lecturer and author, with whom she had studied composition, once advised her to give herself full freedom of expression for her feelings and imagination within the framework of her factual data:

> *Naturally you are scientific. You stay with that and you write a bird book--a guide or studies that interest bird people only. On the other hand you make a book about birds and instead of interesting bird people you interest people in birds. That's literature.*
>
> *A study of nestlings, etc., is only incidental to a larger human meaning--the personal joy, the place in our life, the poetry and meaning of it all from the human, not the factual, point of view. That is the difference between Burroughs and Mr. Chapman.*

In matters of style or technique, however, Cordie was never wholly in accord with Professor Sharp. He helped her far less in finding herself as a literary person than did Dr. Bailey, who always had a direct way of going to the heart of the matter and meeting her need. Although Professor Sharp was correct about the most effective form for her bird stories, it took the blunt kindliness of Dr. Bailey, her mentor in all important decisions, to help her build the style that best served her purposes. Only the long years of tested friendship could lead to such penetrating but constructive criticism as this:

> *"I am not arguing with you, I am telling you," as Whistler used to say. The most important part of a sentence is its beginning, and the next most important part is its ending. Those who write the best English (which is not always the most popular English in these days) are in the habit of starting their sentences with the significant words.*

Dr. Bailey continued more specifically:

> *I was happy to be remembered with your very attractive article about the chickadee. I like "Birdsacre" as a name for a bird sanctuary and home. Your writing is constantly improving. Will you pardon an old friend if he makes one suggestion, however? The most important parts of a sentence are its head and its tail, so to speak. The subject of the sentence ought to be grasped at the*

outset and naturally the conclusion should come at the end. I notice in reading your article that you were rather inclined to obscure your conclusion or weaken it. For example you say "He always announces his name in a gentle confiding cheerful way that is entirely captivating." That is a good statement but the whole force of it is lost by your adding "as soon as he has the slightest opportunity." Now suppose you had tucked that modifying phrase in after the subject, see how much stronger the sentence would have been: "He always announces his name, as soon as he has the slightest opportunity, in a gentle confiding cheerful way that is entirely captivating." You see the announcing of his name is the captivating fact and therefore those significant phrases should constitute the head and tail of the sentence.

Take another illustration:

"He eats the eggs of the moths that would destroy our fruit and foliage when he finds them hidden in the crevices of the bark and around the tender buds, and I have seen him attack the leathery cocoon of the Polyphemus moth on a frosty winter morning, tear a good sized hole in it, and feast on the larva hidden so securely within."

In the first place, the larva wasn't hidden securely *for he* got *it! But that is not what I wanted to say. I want you to transpose your phrase so that the sentence reads this way: "He eats the moth eggs hidden in the crevices of the bark and around the tender buds, eggs of moths that would destroy our fruit and foliage. I have seen him on a frosty winter morning attack even the great leathery cocoon of the Polyphemus moth, tear a good sized hole in it, and feast on the larva within."*

They say any man will pass as a gentleman who has a fine hat and polished shoes. In other words the two ends of him are attractive. The same is true of a sentence.

Criticism such as this led to increased clarity, vividness and vitality. Cordie was a good teacher herself, but she was also an apt pupil, particularly of Dr. Bailey, always a source of sound, practical advice.

Professor Sharp had spoken of the "larger human meaning" in bird study, "the personal joy, the place in our life, the poetry and the meaning of it all from the human, not the factual point of view," as distinguishing the literary approach of John Burroughs. Indeed, Burroughs himself appears as one of Cordie's literary mentors. Her admiration of the great naturalist went all the way back to the Poughkeepsie days when he gave generously of his time and knowledge on impromptu nature walks with groups of Vassar girls. She had shared the pleasure of some of those walks and had been deeply impressed by this kindly bearded naturalist with the soul of an artist. Through the years she maintained her interest in his work and even became an active member of the Burroughs Nature Club in 1916--a "Club of One" as she wryly stated, since the Ellsworth Chapter had never seen fit to ask her to become one of them.

In three letters preserved during her fifty years of correspondence with ornithologists one may see the handwriting of Burroughs. This very busy man found time to offer a candid critique to one who was struggling to express herself. He wrote:

> *Your manuscripts with photos and diagrams are received. Both my son and I have read your article and we agree that the material is valuable but the style of writing is poor. It is jerky and mechanical, no flow in it, no life and vivacity. The paper is loaded down with mere facts and figures. Summarize your results more and aim at more flexibility of style and pictorial expression. Put yourself in it. Easy advice to give, but not so easy to follow. Study and observe the birds not as a matter of business, but for the love of them.*
>
> *You are too bent on making an article. Put more feeling and playfulness in your work.*
>
> *I envy you all those nests. I never found so many of those kinds in all my life.*
>
> *I am sorry I cannot help you, but nobody helped me.*

At first glance this might seem to be overly blunt, but honest criticism must be direct and free of emotion if it is to have value. The heart of the man shows, however, when he says, "Easy advice to give, but not so easy to follow,"--in other words, "I have been through this myself and I do know just what you are feeling now." And then that final gentle push to face up to her own personal challenge: "I am sorry I cannot help you, but nobody helped me."

Men like Burroughs and Bailey demanded the best from Miss Stanwood. They sensed what she had to offer and would not allow her to fail. Mr. Burroughs asked his son Julian to write to Cordelia, elaborating his own opinion about her writing. Julian's letter was equally frank and sincere, one more milestone on the arduous road to effective authorship:

> *My father has asked me to write to you and tell you not to feel discouraged. You have the right spirit and good material and you should not feel discouraged in the least. You must simply digest or work over in your mind what you want to say and then say it just as you would write a letter for instance. And remember O. Henry's rules for writing a story:*
>
> *Rule One---write a story that pleases yourself.*
>
> *Rule Two---there is no second rule.*
>
> *The interest for the public lies in what* you *found of interest in the bird. As long as you interest yourself you will more than likely interest your reader. And the reader has a right to ask that it be written easily so that the style or lack of it does not get between him and the substance.*

Once, while teaching in the Quincy School at Poughkeepsie, Cordie had tried to persuade Mr. Burroughs to talk to her pupils. Although he declined her request, he did write a letter in which he explained his decision and gently chided her: "I should like to talk with them and walk *with them*, but I have no heart for lecturing them. I am sorry you have set your heart on my coming. You must not give your heart to uncertainties."

Actually it was Cordelia's insatiable pursuit of perfection which got in the way of successful writing until she learned to curb her ambition and allow instinct and feeling a free hand. After all, her bird notes overflowed with more good material than she could ever hope to use. Once she became less "bent on making an article," as Mr. Burroughs put it, she began to submerge herself within the tug and flow of those emotional and spiritual tides which had swept her on from one rewarding experience to another. Some thirty-nine individual studies of the most intimate phases of home life among birds, plus general nature essays and special papers appeared in the Audubon Society's *Bird-Lore*, and such other periodicals as *Blue Bird*, *Nature and Culture*, *Nature Magazine*, *The House Beautiful*, *The Wilson Bulletin*, and *The Auk* of the American Ornithologists' Union, and made the name Stanwood synonymous with sound and authoritative field practice made colorful and dynamic by the warmth of a sympathetic heart. In later years she assembled a thick volume of her nature studies, still unpublished.

Leading birders who came to know Cordelia through her articles would repeatedly seek from her additional facts and bird pictures to supplement their own contributions to ornithological literature. Sooner or later somewhere along the way in their correspondence many of them would say in effect: "You are fortunate to be able to study the birds closely under such pleasant circumstances."

Actually, the opportunity for such study is there for all who are willing to apply themselves. Few, indeed, ever do apply themselves, and no birder is particularly favored or fortunate except through the seeking. As to pleasant circumstances, this depends on how well or ill one can take aching muscles after climbing over rough ground, or sunburn, smarting insect bites, and wet feet. Just what favorable circumstance is evident in a steamy, water-soaked swamp, a brush tangle which can pull the hair right out of one's head, or blisters from wearing rubber boots? Serious nature study is no pastime for the timid, the squeamish, or the dilettante. It gives much but it takes much.

When someone told Cordie how fortunate she was, she knew instinctively that in some measure that person had missed the trail which might lead to the inner realm of understanding behind that fringe of trees which borders the conventional approach to nature study. Although that trail is open to everyone, not many discover it, and even those who find it often fail to walk its length.

What she accomplished in this respect was clearly recognized by Dr. Bailey in a letter he wrote on June 5, 1913:

I cannot refrain from writing you a word of apprecia-

tion for your admirable article on "The Hermit Thrush at Home," which I have just discovered in the May number of "Nature and Culture." What a vast amount of painstaking critical observation that article implies. I congratulate you heartily upon being able to observe so closely and to write so well.

He in no way minimized or overlooked the rigors of her labors. He simply paid tribute to the end result of her efforts. This she could honestly appreciate whereas she felt irritated and impatient with envious praise from those who erroneously claimed that they were less favored in opportunity than herself.

Almost any of Cordelia Stanwood's published stories would be suitable to illustrate the quality of her writing and the depth of her dedication, but the two chosen for this book reveal the patient perseverance and sympathetic care with which she approached her subjects and are also particularly appropriate to the Maine scene.

Among the hundreds of nesting bird families studied by Miss Stanwood none was more common or more dear to her than the vivacious black-capped chickadee. Since the chickadee is Maine's state bird its story symbolizes both her home state and her own personal fondness for this "scrap of valor" in feathers. Here is her revised version of an article that had appeared in *The House Beautiful* in October 1920, illustrated with her now-famous chickadee photographs. Her revision reflects the helpful criticism of Dr. Bailey.

TENANTS OF BIRDSACRE

Of the many and varied tenants that lease a building-site at Birdsacre from year to year, no one of them is more beloved than the cheerful, little black-capped chickadee. The chickadee is the first love of all beginners in bird-study. The very fact that he is one of our earliest attachments in the bird-world endears him to us the more since it is proverbial that man loves his old friends best.

The popularity of chickadee is due in great degree to his accessibility. He remains with us the year round. He never changes his garb in youth or age, summer or winter, and he always announces his name, as soon as he has the slightest opportunity, in a gentle, confiding cheerful way that is entirely captivating.

> *"piped a tiny voice hard by*
> *Gay and polite, a cheerful cry*
> *Chic-chicadeedee!"*

What more could the most obtuse novice in bird-lore demand of any bird?

During the bitter winter weather, a bit of suet draws this animated "scrap of valor" to the lilac several times a day where his joyful chatter "Suet-for-me! Chickadee, dee, dee!" gladdens many a prosy hour. But chickadee does more than cheer the landscape with his plump presence, and brighten one's mental atmosphere with his pretty refrains. Chickadee renders valuable service these winter days. He eats the moth eggs hidden in the crevices of the bark and around the tender buds, eggs of moths that would destroy our fruit and foliage. I have seen him on a frosty winter morning attack even the great leathery cocoon of the Polyphemus moth, tear a good sized hole in it, and feast on the larva within. The mite of a bird that looks so helpless in the midst of a vast world of snow and ice, is doing the work it would take a wise man to accomplish, and he sings while he labors! How much we are indebted to this little friend in feathers!

Last spring a pair of chickadees continued to haunt the lilacs, the honeysuckle and the rosebushes around the house. On a morning in early May I heard one of them chittering so persistently, I glanced out the window of Birdsacre Cottage just in time to see a plump little bird with gray back, white wing-bars, black cap and chin, and white underparts tinged with the palest of buffy tints on the sides, busily shredding a bit of cotton wadding. This the bird held down with her feet while tearing off fragments with her bill until the fluffy mass bulged out all around her beak.

I say her, because the female bird is nearly always more gentle than the male bird, and rather more anxious to get the cradle ready for the pretty cream-white eggs with the minute brown freckles all over them. This jolly chickadee never ceased calling to her mate as she picked the cotton; therefore, I knew it must be Mrs. Chickadee. When she had gathered all her beak would hold, she flew to an apple tree near, and laid the cotton on a limb, where she again shredded it carefully. Would that cotton ever be sufficiently fine and fluffy to furnish her dainty abode?

By this time the mate was fluttering in and out of the apple tree, also, and the two twittering birds, the female bearing the cotton in her beak, flew away in the direction of some gray birches that grew along an old stone wall above the cottage. There they disappeared.

Instantly I dropped my work, opened the door, and stole away in the wake of the birds. Before many minutes I had reached the wall and was greatly elated to quickly discover a dilapidated birch-stub that I thought might be the abode of the wee, brave homesteaders. Upon close inspection, however, it proved to have been occupied some previous spring. Nothing daunted, I continued along the wall and a few minutes later came upon another stump, not unlike the first. The top of the bole was about five feet above the ground and slanted toward the rising sun. In the strong morning light, I could see that the birds had hollowed it to the depth of about ten inches, and were lining it with fern-down, fine hair, and some moss. The tuft of cotton batting had been dropped into the bottom, and was ready to be pressed into place, but the birds were nowhere around. They had undoubtedly gone for more lining stuffs.

For a number of years I had been planning to photograph some young chickadees. Here was a nest so charmingly situated that it looked as if it had been made to order! The old, weathered stump against the lichen-covered wall, presided over by two merry chickadees was indeed a subject worthy of the camera. So, very happy, I returned to my work, but said nothing of the pretty secret the old stub by the stone wall was guarding so carefully.

I had found chickadees nesting so many times, I knew about what their plans were. It was necessary to visit them only occasionally to keep in touch with their home affairs.

The nineteenth day of May, I peeped into the stub, but at first could discern nothing. The cavity was so small at the base I began to doubt that the birds were really occupying it. As I grew accustomed to the dim light, however, I discerned a bright eye regarding me

Chickadee nest hole in the old stub by the stone wall. The family raised in this nest is the subject of Cordie's "Tenants of Birdsacre," in The House Beautiful, *October 1920. (CJS photo)*

intently. The brooding bird filled the nest so completely that she held her tail flat against the wall of the hutch. It resembled a bit of weathered stick! The bird was incubating! Although I had visited the nest before, I never had seen an egg. The chickadees are so cautious that the female, after laying an egg, draws the lining materials over it carefully before leaving the nest. And not content with these precautions, although the nest is in a deep, dim hole in an old weathered stump, the last spot in the world where one would think of looking for a softly cushioned seraglio, the careful parent birds begin to brood soon after the first egg is laid, each taking his turn at the task. The chickadee mates share all burdens and pleasures as nearly equally as birds can.

Such rainy weather followed! Two days at a time, with only a break of a few sunny hours, the rain fell in sheets. Always I found one of them shielding the precious eggs without so much as a leaf to protect them from the storm. Could the chickadees survive this constant exposure to the severe weather, I wondered? Even if they lived, would the eggs keep dry and warm? The stump slanted to the southeast, the direction from which the storms came. I feared the small house would be flooded.

The last day of May, I happened to think of the lodge by the wall, and wandered in that direction. Seating myself on a granite boulder near the stub, I waited. Soon I heard "Phee-bee!" Then the sweet whistle drew nearer--"Phee-bee!" A faint, answering "Phee-bee!" and the waiting mate perched on the rim of the nest with a beakful of caterpillars and crane-flies. The

Parent black-capped chickadee bringing a meal of crane flies to the young birds in the nest-hole. (CJS photo)

sitting bird, after partaking of the dainties, flew away, and the fresh bird took up her duties. Before long, the female called her mate from a nearby apple tree, and both birds flitted away for a few moments.

I took this opportunity to peer into the stub. To my great joy I saw four limp, pale-orange chicks about as large as bumblebees. The faithful parent chickadees had won out!

For the first time I was able to determine when I should be able to photograph the little family, if no foes discovered them. Young chickadees remain in the nest about fifteen days. In two weeks the nestlings would be large enough to have their photographs taken. How pleased the boy who helps me take pictures would be when he learned of the pretty family in the old stump!

At last the time was propitious! The birds were so gentle, the boy and I placed the camera near the nest without any screen to conceal it, but as the sun remained on the stub but a few minutes, we had to try several mornings before we secured prints of the mature birds. The boy and I were kept very busy placing the youngsters. As soon as we posed them in focus, one or both parent birds returned with immense caterpillars and moths in their beaks, and perching in trees behind and above the young birds, called "Chickadee, dee, dee!" vigorously, causing the youngsters to twist their heads around, flap their wings, and flutter off the perch into the grass in every direction, like so many bits of thistledown. The boy found them and returned them to the branch, I do not know how many times. It was

Six little chickadees sitting for their portrait. (CJS photo)

The one baby chickadee who "acted as if he would like to pose forever." (CJS photo)

also his duty to determine which babe posed most successfully that we might catch his likeness in different positions. At last, he selected one, a pert little chickadee who acted as if he would like to pose forever. Him, we snapped again and again. By the time we had exposed the last plate, most of the babes were very sleepy, and upon being put back into the nest, snuggled down blissfully, but the pert little chickadee hopped out of the nest as fast as we put him back. At last we decided to let him sit on a branch over the stump. He remained very quiet until a parent chickadee appeared with craneflies, when he flew up into the tree to meet him, and was lost in the foliage. I knew that by the following day the whole family would be on the wing.

The chickadees continued to twitter and gather food in the shrubbery round Birdsacre Cottage all summer.

Next to the vivacious and carefree black-capped chickadee no small bird is more captivating than the jolly, bouncy, zany, "upside-down" red-breasted nuthatch. His ribald "quank! quank!" seems like a happy Bronx cheer expressing casual indifference toward the world at large, shouting in boisterous good nature: "You go your way and I'll go mine."

Cordelia found the cheery chickadee and the bold nuthatch her constant companions to brighten any day on the forest trails. Since both were resident birds she could enjoy this friendship in all seasons through fair weather or foul. It is doubtful that anyone has enjoyed a more constant or happy intimacy with birds than Cordie knew with her chickadees and nuthatches. Among her papers at Birdsacre is an article that captures this enthusiasm--a story about red-breasted nuthatches written in 1912:

THE "UPSIDE-DOWN" BIRD

The red-breasted nuthatch is a quaint-looking beautifully marked bird and well adapted, because of his long beak and strong feet, to pasture on the bark of trees. Although the nuthatch makes a specialty of caring for the bark of trees, he leaves no source of revenue untapped. Sometimes he gathers insects from the foliage, particularly when the spruce bud moths are common. Occasionally he extricates seeds from the cones and eats them, or hides them in the crannies of the bark for a day of need. He even descends to the ground and forages there, not forgetting to investigate thoroughly each old, mossy stump and log.

Though the red-breasted nuthatch is common throughout North America he returns to the coniferous forests to build his nest and rear his young. Every spring and fall he calls more or less generally throughout the evergreen and mixed growths, and even from the trees of busy thoroughfares. The call note of the little bird is so inconspicuous that he might spend many hours in the elms without his presence being discovered, but a bird lover always hears his nasal "quank! quank!" and cannot refrain from stopping to watch the little gymnast walking head-foremost down the tree as deftly as he ascends it, or out the underside of a branch with as little effort as he trips along the top of it, all the while dexterously removing noxious insects from the cracks and crevices in the bark.

While the nuthatch is industrious, he loves company and after his nesting duties are over he is never too busy to utter his "quank! quank!" to inform his companions of his whereabouts. One fall I came upon a large band of these cheerful gypsies in their favorite forest. They called back and forth constantly. The effect was not unlike a band of hylas in full chorus. If a nuthatch strays from his kind, he will frequently attach himself to a company of brown creepers, golden-crowned kinglets, or chickadees. To see one of these society loving birds is to be sure that a troop of foes to insect pests is in the neighborhood.

One precious tract of ancient woods, bedecked with lichens, overgrown with moss, and, alas, now fast disappearing beneath the axe of the woodsman, frequently harbors a pair or two of red-breasted nuthatches during the breeding season, but only rarely do they nest in this vicinity in large numbers, and it is but occasionally that the nuthatch remains to partake of my suet when the snowdrifts lie deep and Jack Frost sketches on the windowpane.

In the year 1912, however, the red-breasted nuthatch did winter among us and in the spring the number of those that tarried in our midst were augmented by fresh arrivals. They nested everywhere--in dead stubs, in trees in the forest, even in a dead tree opposite one of the city grammar schools. Its nest resembles that of the chickadee and the woodpecker--a chamber excavated in a tree. In my locality the bird chooses to fashion his domicile in a dead fir or poplar, the wood of which is soft enough to be easily worked by the beak. They start the little door much as if they were trying to bore to the cell of some hidden grub, and it is very possible that in looking for food they do come upon suitable building sites.

One pair spent about eight days in hollowing a nest which I afterward examined. Both birds toiled very nearly the same length of time in shaping the nesting space, fashioning the cavity and removing the waste wood with the beak. The male threw his chips from the door of the little house most vehemently, frequently coming out to pick off any fragments of wood that clung to the bark. At such times he would dart swiftly back again to his labor as if it were imperative that he should complete a certain amount of his task before he lost his turn at home building. While the female was awaiting her chance to display her skill as a carpenter, she often stood quivering on a branch overlooking the nest. Again she would flutter on the bark outside the door, or at times she would become so impatient that she slipped into one or the other of the two holes excavated above the nest and pecked out a number of beakfuls of chips while she waited. These, as was her habit, she cast abroad from the branches of neighboring trees.

While the birds were hollowing their living room, they appeared not to notice my presence. I sat on a log not far from the nest and came and went when I chose, but the moment they began to bring shreds of yellow birch bark with which to line the cavity they became very wary. Every time I approached the stub, they ceased all work, and refused to return no matter how long I concealed myself in the surrounding seedlings. Lining the nest occupied about five days.

After the tiny tenement was entirely fitted up they smeared the space around the entrance with pitch; more was placed at the bottom of the little door and at the top and sides. It ran down the bark when fresh in small globules. Every few days as the pitch dried into the bark it was renewed. I suppose that the birds did this to ward off some enemy, perhaps ants. I have never found insects or parasites in the nest of the red-breasted nuthatch such as one finds in the nest of the chickadee, the downy woodpecker, or the flicker, nor have I ever known of a nest of nuthatches to be robbed by the wild folk of the woods.

In due time the eggs were laid, the female brooded, and the male waited on her assiduously. When I had a leisure hour, I spent it watching and listening to the nuthatches. They were gentle and fearless, but extremely cautious birds.

Most of his time the male spent foraging for food. He fed his mate regularly with grubs, caterpillars, crane flies, and moths. As he drew nearer the nest stub he called "quank" from time to time. At last, he fluttered down to the door and delivered the food to his mate who stood in the doorway while receiving the tidbits. As he flew away, the male twittered "que! que! que!". When the fair one permitted, he escorted her to the woodlands, or some great lone pine or spruce in the "Cut Down" for food and exercise.

After the little mother had been brooding some

fourteen days, I saw her carry from the nest what appeared to be a bit of egg shell. That day the male fed her as usual, and she apparently cared for the little ones. The following day, however, the male entered the nest every other time that he came with food, and ministered to the young birds. Occasionally the female lingered for him to deliver the food to her when he tried to feed the young, but he refused to do so. Once when the male remained in the nest some time, the female came to the window and twittered as if to say: "My dear, do you realize how young and weak our babies are?" He quickly withdrew.

It was most unsatisfactory to sit on a log day after day and wonder how many nestlings the birds were feeding and just how little nuthatches looked. At last I could curb my curiosity no longer, so the boy and I carefully sawed off the upper part of the stub, tied it to the lower part, and removed a section below the nest hole. Six small, blind nuthatches were snuggled down in the warm bark cradle. I avoided keeping the nest open for more than a few seconds for fear of lowering the temperature.

When I began to examine the stub, the parent birds disappeared. The following morning, however, the female was in the nest and the nest hole was newly smeared with pitch. A few fragments of feathers clung to the balsam.

When the young were thirteen days old, they were well feathered and resembled the parent birds closely. That morning I carried the stub containing the young to the studio. I was able to feed a little bread and milk to the least mature nestlings, but the others refused to eat.

Only a few moments after I tied the nesting stub in place in the woods again, the parent birds appeared in a spruce above the nest and twittered softly to each other while they examined the nest hole from different angles. The famished young became very restless and called "pit! pit! pit!" a number of times. But the moment the parent birds said "quank" once or twice the hungry bantlings became silent. Then the parents went away and although I remained nearly an hour they did not come with food, nor did the babes utter another sound.

When the young were sixteen days old, I visited the nest again. I tapped gently on the bark and they came to the window expecting food, but not getting any they twittered and then three of them shot from the door one after another calling "quank! quank! quank!" They easily flew five yards.

Stub with nest of the red-breasted nuthatch, cut open to show the contents.

Then ensued a charming lesson in bird craft. The mother came quickly, a caterpillar in her beak, alighted beside a cute little nuthatch on a bole, and then crept up the trunk ahead of him. The young bird followed and was rewarded with the tid-bit. Then she alighted beside a fledgling on the ground and persuaded him to run up a tree trunk. The young bird ran up a few inches and clung there. She fed him and he continued up and vanished among the branches. In like manner she enticed the third to a place of safety. A red squirrel trilled angrily from the top of a stump.

The first photographs were not a success so two days later I took the remaining three nuthatches to the studio once more. They had grown so strong and were so active that the dainty little nest of yellow birch bark which had made such a comfortable cradle for the newly hatched young had been ground to dust.

One of the little nuthatches while posing flew and was lost in the studio. Although we searched until it was time to close the rooms we found no trace of him. It was 6:00 o'clock when I returned the nest stub with its two remaining occupants to its proper place in the woodlands. The young birds called and scolded, and although I waited until it was quite dark, the parent birds did not appear. Fortunately, the fledglings remained in the nest.

Twice that day I had been caught in thundershowers and wet through. I had walked all of ten miles, and I had yet to cross a "Cut-off," a wooded swamp, and a dense growth of timber before I made the woodroad that led to the street and home. I should not have minded any of the hardships of the day in the least if I could have been sure that all of the birdlings were entirely safe.

The next morning when I called at the studio the lost nuthatch had been found and put in a well ventilated box. Weak and nervous, the baby bird called constantly, but refused to eat. I promptly set out for the nest site with the box and the baby bird seemed to know where he was the moment I entered the woodland. His cries became longer and more insistent. They reached out as if to compel the parent birds to come to his rescue.

At last, although in my anxiety it seemed as if I never should reach the home site, I placed the baby bird in the nest stub. The other two had already flown away.

Would the parent birds miss the sixth nestling and return his call?

The little fellow continued to send out long, pitiful, pleading calls. Voiced in words they surely said:

The baby red-breasted nuthatch lost in the studio and then found and returned to his family.

"I am so ill, so hungry. I need my mother sadly. O, Mother, dear mother, please, please come!"

Almost before I could remove my hands from the nesting stub the mother bird flew down toward the nest twittering sweetly and softly and the hungry youngster darted from the door to meet her. He had been without food for nearly twenty-four hours. They flew away together.

Now that I had delivered the last little nuthatch safely to his parents, I was so relieved that I was quite content to sit on the log beside the now deserted nuthatch abode and just rest.

The last of the six little nuthatches had left this snug homesite with his mother on his eighteenth birthday.

Tired as I was I thrilled with happiness for this reunited family, and felt deeply grateful to them for the privilege of sharing their home life so intimately.

9

Recognition

The reward of a thing well done, is to have done it.
Emerson

When Ora Willis Knight's *Birds of Maine* appeared in 1908, Cordelia Stanwood had already outgrown her amateur status and was writing professional papers on nest life. Others who were building the bird literature of the northeastern United States began to solicit even more of her patient and meticulous observation. Still only a few were engaged in this groundwork essential to a more thorough understanding of the life of Maine birds.

She bought a copy of Knight's book in 1909 and annotated it with her own observations and deductions. Although only twenty miles separated his home city of Bangor from her own Birdsacre she knew him only through correspondence. This was typical of Cordie's contacts and collaboration with other naturalists. Many of them lived in distant areas of the country, one even in England, and few ever reached out-of-the-way Maine in their travels. Cordelia preferred it this way. She had always been a solitary worker and would never be quite at ease when directly involved with others.

One of the first important birdmen to "discover" Cordelia Stanwood and to recognize the value of her work was Frank M. Chapman, the man whose *Birds of Eastern North America* had helped her so much in the early days of her apprenticeship. As early as 1908 while editor of *Bird-*

Lore, the official organ of the Audubon Societies, he wrote to her:

> *I have examined with interest your notes on the "Bay-Breasted Warbler." They seem to me to constitute an addition to our knowledge of the life history of this species, well deserving publication. I should, however, make some slight modification in the manner of presentation; chiefly in your use of quotation marks, the significance of which is not to me always clear.*
>
> *If I were you, I should omit the quotation marks altogether, and put the paper in the form of a summary of observations with dates where they are needed. Anything that would result in condensation or abridgement would also be desirable; for a scientific communication of this kind the facts should not be concealed by unnecessary or unimportant statements. I will add, however, that there seems to be opportunity for but little abridgement in this paper. Your months of observation must have placed you in possession of much valuable material, and I wish it were possible for me to discuss with you methods of presentation.*

Dr. Bailey's long residence in Ohio as Director of the Cleveland School of Art had bearing on Cordie's contacts with the bird people of the mid-west. First among these was Dr. Eugene Swope, Field Agent for the Audubon Society of Ohio with offices in Cincinnati. Cordie was one of the earliest writers for *Nature and Culture*, edited by Dr. Swope, and later for *Blue Bird*, its successor. An ardent and tireless advocate for nature study and conservation, Dr. Swope was the kind of man not at all dismayed by a lecture tour in Florida in 1914 which brought him before two hundred and ninety-five audiences. Always pleased to receive material from Miss Stanwood of Maine, he paid her a tribute both wry and sincere: "You are a good and prolific writer and I wish I had the means to pay you better for your articles."

Eventually, Dr. Francis Herrick, who was doing intensive research on nest construction, learned about Miss Stanwood's work and requested actual nests from her for the museum of Western Reserve College in Cleveland, Ohio. She met this request with perfect nests ranging from the red-

Cordie became an international authority on nests and nesting. In December 1939 Nature Magazine *featured a collection of her nest photographs. The caption for this nest of a chestnut-sided warbler, said: "Where the tent caterpillar spreads its structures, the chestnut-sided warbler binds her dainty cradle to the stems of meadowsweet bushes. Sometimes the outside of the nest is shaped of coarse grasses and the lining made of soft, fine hay; at other times the entire structure is fashioned from caterpillar silk and fine hay. The nest of this warbler is usually carefully built." (CJS photo)*

start cradle in a birch fork to the eaves-swallow home of cement-like mud. This collaboration was particularly rewarding to Cordelia, since later, when Professor Herrick traveled abroad in England and France, she enjoyed his colorful letters recounting his family's experiences overseas.

She was quite flattered when Edward A. Armstrong of Cambridge, England, wrote to her and asked for further particulars about the winter wren, related to but somewhat unlike the wrens of the British Isles:

> *I am writing a book on the wren of these islands, which as you will know is related to your winter wren,*

and having written to Dr. Bent after seeing your observations mentioned in his book he gave me your address.

I should be most grateful for some of your notes about this bird. The references which interest me most are on p. 152 of Dr. Bent's book. You refer to the female winter wren building the nest. I should very much like to have details of this. Very occasionally the female of our wren will build. I wonder if the birds you mention were exceptional in this way or whether you have evidence suggesting the female winter wren builds more frequently. Judging by other evidence it is rather hard to believe the male does not often build.

I am very interested, too, in your mention of the male feeding the female. In our bird this is exceedingly rare.

Our bird usually only builds when there is wet material available--during or after rain. Has this been your experience? I find that towards the end of the season the male frequently helps to feed the nestlings.

Our bird is a polygamist. Have you any evidence of this for the winter wren?

If you have notes of the feeding rates, length of time the mother covers the young at different ages, and so forth, I would be most grateful to have them to compare with my own.

My apologies for bombarding you with so many questions but I would indeed be much helped by any light you can throw on the ways of this delightful and fascinating bird.

She sent him detailed records from her notebooks and eventually had this note of appreciation:

Very many thanks for so kindly sending me your wren notes. I have looked through them, though not yet studied them carefully, but I can see that there is much of interest in them. I hope it will be all right if I keep them by me for a little while as I am writing my wren book at the moment and would find it valuable to be able to consult them on various points.

It rather surprises me that on your side the winter wren has not aroused more interest as it must be a very fascinating bird.

When Arthur Cleveland Bent of Taunton, Massachusetts, began the compilation of his monumental *Life Histories of North American Birds* for the Smithsonian Institution, Cordelia Stanwood became one of the many contributors to this great cooperative undertaking. Her elaborate notes, manuscripts, and photographs of the life histories of at least a dozen species found their place in Dr. Bent's work. Cordie's scrapbook holds many letters over a score of years from this dedicated scholar. This kind of communication and friendship meant much to her. It was vindication for what she had achieved, and it kept her alert to the steady growth of ornithological knowledge. She felt great satisfaction to share in and contribute to that expanded knowledge.

Unquestionably, Edward Howe Forbush's three volume *Birds of Massachusetts and Other New England States* is still a definitive work in bird lore for northeastern America. Here again Miss Stanwood has her place, both in text and photography. She is also well represented in the more recent *Maine Birds* by Ralph S. Palmer, a standard text for any student of Maine birds.

In February of 1916 Elizabeth Goodwin Chapin wrote to Cordie from the office of the Burroughs Nature Club in New York:

> *Several of our members who had read your article on the Red-breasted Nuthatch, have written the Nature Bureau (which we maintain for Burroughs Club members) asking why you and John Burroughs disagreed on the point of the nest excavating habits of the Nuthatches. Burroughs says in WAKE ROBIN, page 114, that these birds lack the strength of bill to excavate, and that they must adopt a hole already formed, excavating just enough to alter it. We have quotations from Blanchen, Dugmore, F.M. Chapman, and Elon H. Eaton, all more or less contradictory. Mr. Clinton G. Abbott has written that he knows the bird can dig out rotten wood, but that while he has seen it in the act, he does not know whether or not the hole was begun by the bird's own initiative.*
>
> *Would it be too much to ask for a little letter from you, saying whether the birds you saw began their holes,*

or whether they just improved natural cavities in the decayed wood?

Door of the red-breasted nuthatch hole surrounded by pitch. Photo by Embert C. Osgood for Cordie's article "Craftsmen Cradles and Babes They Have Held."

Miss Stanwood's reply, prompt and emphatic, eliminated any possibility of doubt as to the nest-excavating ability of the red-breasted nuthatch.

Through Miss Chapin's answer Cordelia received another kindly tribute from that great naturalist who had encouraged and inspired her in so many ways over the years:

Mr. Burroughs is in town this week, and comes into the office every morning to get his letters, prowl around among the new books, exchange a little gossip, etc. He is having his picture painted by a Hungarian artist, an able

> *woman in her craft, and a Princess by marriage, with some frills attached to this feature, and we have to joke him a little about his royal connections, how one behaves with a Highness, and other absurdities, all of which he takes very smilingly.*
>
> *I have asked him just now what he thought about nuthatches being able to excavate, and read him Mr. Clinton G. Abbott's letter, stating that he, Mr. A., has seen the birds flying away with chips in their bills. At first Mr. Burroughs said, "I don't believe it." Then I read him your letter, and he said, "Well, I know Miss Stanwood--she is a real bird woman, and if she has seen this trait, I believe absolutely what she says must be true, though personally I have never seen a nuthatch excavating." He added that he never had opportunity to see the Red-breasted variety at nesting time, and felt sure the White-breasted ones he had seen had not excavated beyond removing a little rotten wood from a hole already bored.*
>
> *Since we have had some correspondence on this topic, I thought you would be interested to know what Mr. Burroughs said, and also his very pleasant comment on you as an authority.*

Burroughs was not alone in his judgement of Miss Stanwood and her work. In January of 1921 Dr. Bailey wrote:

> *Dallas Lore Sharp was the guest of Mrs. Bailey and myself at a dinner in his honor here in Cleveland last Saturday night. What do you suppose he said about you? He said that you were doing the best work in bird study that any woman has done in the United States since the days of Olive Thorn Miller. "What do you know about that?" I hope that bit of information will encourage you.*

Olin Sewall Pettingill, Jr. of the American Ornithologists' Union paid tribute in this way:

> *Your information on the birds in the vicinity of Ellsworth has been most useful and I am greatly indebted to you for sending it. I hope that I may meet you some day. Your papers on the wood warblers have been among the most notable contributions to our knowledge of the group.*

Finally, after her extensive contributions to Bent's *Life Histories*, Dr. Alfred O. Gross of Bowdoin, who also worked on this massive study, offered her the ultimate compliment when he said: "Your life histories of birds such as the Hermit Thrush and many others are the best any Maine person has written."

Her portrait is in the Deane collection presented to the Library of Congress by the American Ornithologists' Union in 1934. Three years after her burial in the McKenzie Cemetery in Ellsworth, the following citation was received by The Stanwood Wildlife Foundation:

THE BUREAU OF
SPORT FISHERIES AND WILDLIFE
UNITED STATES
DEPARTMENT OF THE INTERIOR
WASHINGTON, D.C.
Gratefully acknowledges the faithful services of
MISS CORDELIA STANWOOD
in reporting, for use in scientific investigations,
observations on the distribution, migration, and
abundance of North American birds, for 32 years
during the period 1910 to 1946.

10

The Photographer

To him who in the love of Nature holds
Communion with her visible forms, she speaks
A various language . . . .
Bryant

Cordie was quick to recognize the value of good pictures to illustrate the articles and stories for which she was finding a market.

In the early days she would borrow a fledgling from its nest, tuck it snugly under a bit of soft cloth in one of her many baskets and walk the two miles to town to have Embert Osgood, the Ellsworth photographer, take its picture. This arrangement brought no great profit to Mr. Osgood,

Three young red-eyed vireos and their nest--one of the photographs made for Cordie in the studio of Ellsworth photographer Embert C. Osgood, before 1916.

but he was sympathetic to Miss Stanwood's dedication and found ample reward in observing her delight over a good picture.

Afterward she would hurry back to the nest with her young charge and be soundly scolded by the distraught parents for disrupting the tranquility of their domestic arrangements. She always felt acutely guilty for disturbing the natural order of events among her feathered subjects, but fortunately, she consoled herself, it led to no serious harm, and it did help to deepen and expand her understanding.

Before long Cordie came to realize how much better it would be to have her own equipment and take pictures directly in the field. Her share of Aunt Cordelia's estate in 1916 gave her the means for such an investment. But what kind of camera should she have and where would she find it? No other than Frank M. Chapman, with whom she already had correspondence, came forward to help her. Mr. Chapman, director of the American Museum of Natural History and one of the original editors of *Bird Lore*, the Audubon Society's first magazine, had accepted several of her bird stories, and his books on the confusing warblers had been most helpful in preparing her for her own contributions to their study. Mr. Chapman was himself an accomplished photographer, and through his efforts she acquired a large, box-type Eastman Kodak camera which used 5x7 glass plates and was mounted on a sturdy wooden tripod.

With great excitement she quickly learned, under the guidance of Mr. Osgood, to operate the bulky, complicated equipment, and was soon producing prints of professional quality, especially after she invested $122 in a telephoto lens.

In the decade between 1916 and 1926 Cordelia Stanwood took well over one thousand pictures, mostly of birds and related nature subjects. Some of them, however, were studies of historical homes in Hancock County, including the antique furniture, glassware, and other treasures found in them. *The House Beautiful* published her illustrated stories about old homes in Blue Hill, Castine, and Ellsworth.

Her black and white studies of birds, nests, and flowering shrubs added intimate vitality to her published articles. At first she sent her exposed plates to the Kodak company

Hobblebush (Viburnum alnifolium), *in blossom and fruit, a common shrub at Birdsacre. (CJS photo)*

for development. Since this was rather expensive, she learned to develop her own and make prints. To have some bird which she had observed closely or nurtured to maturity achieve visual permanence on a sheet of white paper more than justified the tedium of such work.

From this canvas blind Cordie photographed the olive-backed thrush and other bird families. (CJS photo)

Normally, Cordie would not tolerate the presence of others when doing research in woodland and field. True, her isolation was usually well protected by the necessarily long hours of quiet attention, the constant irritation of swarming insects, the sheer drudgery of just waiting for something to happen. Quite accustomed to looking after herself, she gave little thought to cutting poles and carrying brush for the numberless bird blinds which she built. She accepted such labors as a part of the process and enjoyed her work.

But when she went into the woods with her camera she could no longer manage alone. There was too much to carry; the equipment was too cumbersome and heavy. Fortunately, the Stanwood home on Beckwith Hill was always a favorite gathering place for the boys and girls of the neighborhood, and Cordie had never lost her enjoyment of children. Harvey and Raymond Spillane, and after them their younger brother Clarence, became her helpers, as well as her nephew Alfred when he came to visit his grandparents. These boys not only helped in carrying and setting up camera equipment, but were equally agile at climbing trees and building brush blinds which could be reached by ladder.

From brush blinds like this Cordie watched and photographed birds at the nest. (CJS photo)

With her camera, Cordelia added another dimension to her observation and found a new application for her artist's training. Although primarily a mechanical device, the camera appealed to her artistic temperament, and she became proficient in evaluating the significance of composition, lighting, background, and contrast. Her high quality photographs not only illustrated her own published material, but were purchased by other ornithologists such as Forbush and Bent, who recognized their excellence.

On the back of this photograph of a home in Ellsworth, Maine, Cordie has written: "Residence belonging to Mrs. George Dutton. Miss Louise Dutton the novelist who lays the scene for many of novels in Ellsworth was born here. Mr. Hannibal Hamlin 2nd rents it." (CJS photo)

The entrance to "Stanwood," the Bar Harbor cottage of Presidential aspirant James G. Blaine. Mrs. Blaine was a Stanwood of the Augusta branch. (CJS photo)

11

The Arts and Crafts

A picture is not wrought
By hands alone, . . . but by thought.
In the interior life it first must start, . . .
The rest is but the handicraft of Art.
Story

Cordelia Stanwood's life may have appeared simple to a casual observer, but from her childhood she was an extraordinarily complex person, and as she grew older, the complexities expressed themselves in paradoxes.

Frail as a child and subject in early adulthood to recurring headaches that reduced her to near immobility, suffering from afflictions that resisted diagnosis, she nevertheless went on to a rigorous life that required a stamina and endurance that would challenge the physical fitness of the most robust.

Paradoxical in another way was the contrast between her unconventional life pattern and her genteel Victorian upbringing that established in her the high standards of propriety in dress and behavior that were the hallmark of a lady. Outwardly Cordie authenticated the image of the Victorian lady by carrying into the middle of the twentieth century the long black skirts, the high-necked shirtwaists, the piled-up hairdo, and the high-buttoned shoes. But the prim, precise lady who held a teacup so daintily while entertaining friends in the parlor of her home was also the woman who tramped for miles through woods and swamps with hip boots under the long skirt to locate the nest of that winter wren which had eluded her the day before. Cordelia was a lady, but a lady with a difference--the Lady of Birdsacre.

In a period when most women were content to sit in their parlors and tend to their needlework, Cordie would be out and away across the countryside--undaunted by the rough terrain of the Maine woods. Quietly concealed, she would sit for hours on end, capturing every intimate detail of the nest life of her subjects. Yet paradoxically in this spell-bound immobility she was most vibrantly alive--breathing and feeling the very heart-beat of the life around her.

Another of the paradoxes of Cordie's life was her relationship to the traditional women's arts. In one way, her real home was under the open skies or in the close shelter of a brush blind. The everyday domestic arts of housekeeping appealed to her not at all, and she systematically ignored them or retreated from them as beneath her contempt. But she had been born with a gift of unusual manual dexterity. Her hands were strong and supple, and her fingers deft and swift at any task she put them to. Her needlework was finer, her knitting firmer, her crocheting and embroidery more delicate than that achieved by others whose acquired skills were a routine essential in every girl's upbringing. From the days of her childhood when she had first learned to darn and knit at her grandmother's knee, Cordie was never far removed from some constructive expression of her physical skills. This gift contributed to her success as a drawing instructor and became a major support in the therapy which would eventually bring her back to reality after the breakdown that had destroyed her teaching career and forced her to return to Maine.

And through her hands Cordie found a way to express herself. When she could not relate to people, when she turned away from conventional paths and rebelled against social restraints, the discipline of her handicrafts served to steady her nerves and calm her rebellious spirit. Although not remarkable, her paintings also reflected a sensitive response to the beauty and harmony in nature.

Unfortunately Cordie was even less effective in managing her financial affairs than she was as a housekeeper, and she frequently found it necessary to pinch and save and do without. Somehow she managed, largely through her handicrafts. The sale of innumerable baskets--some sturdy and serviceable, others intricate and ornamental--countless hooked and braided rugs, glove and jewelry boxes and picture frames in exquisite burned woodwork all helped to

In one respect Cordie continued her teaching career through her instructional articles for crafts and homemaking magazines. Here is the illustration for instructions on making a raffia reticule. The original in the photographs is still at Birdsacre.

supplement the limited and uncertain returns from her writing.

In basketry she was not content to construct just the currently popular designs but went to live with the Old Town Indians on the Penobscot while learning to duplicate their age-old craftsmanship in ash and sweet grass. We do not know exactly when she went to Indian Island to study basketry with the original craftsmen in this ancient art (probably between 1910-15), but her remarkable proficiency would indicate that she must have been there for many weeks to have been able to reproduce so many kinds of baskets of every

This acorn knitting basket shows Cordie's mastery of the split ash and sweetgrass basketry techniques of the Penobscot Indians with whom she studied.

size, shape, and design over a period of several years. It is easy to imagine that she, who was more at home in the woodland than inside a house, was at ease among a people who had always lived in harmony with nature.

One is reminded of the innumerable neatly constructed brush blinds Cordelia built over the years which so closely resembled the design of the Indian tepee, a frame of poles lashed together at the top and closely laced with spruce and balsam boughs to form a snug retreat from which she could observe her avian subjects more closely: in effect a kind of over-sized basketry to meet a need by simply utilizing the natural materials at hand.

Later she took up weaving and set up a large loom in one of the upstairs chambers. It literally shook the house when in use. Cordelia had spent most of the summer of 1911 in the little town of North Lovell near the New Hampshire border where she boarded at an old farmstead and studied weaving with Mrs. Cushman Sawyer, who had learned the craft with the highly acclaimed Volk family at Hewnoaks in Center Lovell. This was in sparsely settled farm country around Kezar Lake which Cordie found very much like her own familiar hills and valleys at home. Between lessons she had enjoyed tramping across the fields and pastures to find the birds and flowers common to the area.

All forms of handicraft appealed to Cordie, but she never pursued weaving with quite the same dedication that she applied to basketry. Eventually she gave up her loom to the Seacoast Mission which ministers to the inhabitants of Maine's off-shore islands, and it might well be that it is still in use behind some storm-lashed island shore. Cordie would like that.

Pin tray and covered basket, Cordelia Stanwood's design and execution.

For many years Cordie derived a large part of her income from her basketry. These baskets of her own design and execution demonstrate both her skill and her excellent sense of design.

12

A New England Eccentric

No man is quite sane; each has a
vein of folly in his composition . . . .
Emerson

"For non-conformity the world whips you with its displeasure." Emerson had said this before Cordelia was born, but she had lived to know its truth and to accept it with true Emersonian self-reliance.

One who walks alone invites suspicion. Since she either avoided people or ignored them, she was often judged harshly or rejected for imagined faults, especially after she gave up teaching and returned to Maine. Cordie's living apart from the conventional patterns of behavior led to the assumption that she was somehow peculiar.

For those who chose to discover a strong thread of disillusionment and unhappiness in the fabric of Cordelia Stanwood's life it was all too easy to interpret her aloofness, her unorthodox behavior, and the austerity of her hermit-like existence for over twenty years as conclusive evidence of a troubled spirit who had missed much of the joy of living. The near poverty in her later years so sharply opposite to the advantages and privileges that were hers in early life might seem to support this assumption.

But the truth of the matter is quite different. Neither disillusioned nor unhappy, Cordie had simply made herself indifferent to outside influences. Unlike most, she had not outgrown or lost her youthful dreams. She just took them with her and always found her happiest moments when least concerned with mundane things. She seemed to function in a

One of three snapshots of Cordie in her later years in Ellsworth. The photographer was almost certainly Ida Wilbur.

world of her own. Many stories about her unorthodox way of life circulated in the community, stories which acquired the aura of local legend after countless repetition by those whose imagination suffered no restraint in the telling.

The roots of Cordie's eccentricity appear early in her family relationships. Not long after she came home in 1904, and while her brother was still living with the family, Cordie came downstairs one morning after Harry had been celebrating with friends and found him, still fully clothed and noisily asleep, sprawled across the love seat in the parlor. Saying nothing to anyone, she walked out to the well, drew a pail of water, went back to the parlor, and emptied the pail over Harry. Her answer to her mother for what she had done was simple and direct: "Harry was hot and uncomfortable and needed to be cooled off."

It is to be noted that she did not say that Harry was drunk, that he had no business to be in the parlor in that condition, or that as a member in good standing of Roger Williams' First Baptist Church of Providence she just could not stand such conduct. It matters little whether it was a glassful or a pail of water which she had poured over her brother. What signifies is her positive action, when, to her way of thinking, such action was called for. Not for a moment did it cross her mind that others might consider her conduct childish or odd.

The water spilling incident reveals a growing lack of communication and understanding between Cordie and her brother that rapidly worsened after several clashes over household management and uncompromising differences in personality. Cordie was the oldest of the five children, Harry the youngest. The only boy in the family and spoiled by an indulgent mother, Harry was rather selfish and demanding. Always painfully correct and proper in her own reserved way with other people, Cordie found it very difficult to accept or ignore her brother's slipshod ways. Soon after his marriage in 1911 Harry left home and thereafter neither he nor Cordie ever made any effort to bridge the gap between them.

When Cordelia had gone to Springfield, Massachusetts, in 1894 as assistant supervisor of drawing, she had found that one of her eighth grade art pupils was a rather sober little girl named Georgia Jordan. Georgia's grandparents had built a house on the Buttermilk Road in Ellsworth less than a half mile from the Stanwood home. Georgia took Cordie back to her own childhood, recalling her visit to the Jordans with her own grandmother when they had lived at the foot of Beckwith Hill. She recalled going to the damp banks of Card's Brook to collect mosses with Miss Sarah, an aunt of this child who now came to her across the years. How small the world and how infinitely fine the threads that bind one life to another!

Georgia Jordan Ray always remembered her art teacher with fondness, and after Miss Stanwood's death when she herself had become an elderly widow said that she "was very beautiful and dignified and we all loved her. I never knew her to get upset about anything, and she had more patience than anyone I ever knew."

Georgia Jordan's son, Frank Ray, was a fish peddler in the Ellsworth area in later years after Miss Stanwood had come home. Sometimes when returning at the end of the day, he would stop at the house on the hill and offer a fillet or a whole cod that had not been sold. Cordie always firmly declined the gift. She would be gracious about it, but she was too proud to accept anything unless she, in turn, could offer something of equal value. This pride would both sustain and hamper her in those long lean years ahead when she would struggle on alone to interpret and perpetuate the woodland sanctuary that was her home.

Nevertheless, Frank Ray was also capable of Yankee stubbornness. He met Stanwood pride with his own brand of homely ingenuity. Unobtrusively he would slip the wrapped package into the mailbox and go merrily on his way. Acceptance became a matter of expediency since such perishable products could not be left in a mailbox, nor could Cordie with her Baptist background just discard this offering and be guilty of sinful waste.

At other times, however, Cordelia was much more stubborn when someone offered a helping hand, particularly in the later years when she was alone and had so little to provide even the bare necessities. Her sister Della tells about one occasion when she had come home for a visit and had been disturbed to discover that Cordie was becoming more and more uncommunicative and withdrawn. For one thing, she was not eating properly, relying too much on crackers, canned milk and the like. Without saying anything to her sister about it, Della made arrangements with a local dairyman to have a quart of fresh milk left on the doorstep each morning.

Cordie was infuriated. No one, not even her sister, could intrude upon her closely guarded sense of privacy and independence. She simply ignored the deliveries. Since it was wintertime and milk in those days was distributed in glass bottles, frozen milk and broken bottles dramatically emphasized her displeasure.

Another incident involved a cookstove. For the twenty-five years that Cordelia lived alone after her mother's death in 1932, she abandoned the lower floor of the homestead and chose to live as a recluse in the upstairs room where she had been born. Up under the shingled roof in the middle of the house she apparently felt more secure. That all of her firewood and water had to be carried up a treacherous flight of stairs from the dining room seemed not to bother her in the slightest. The old cookstove in the unused kitchen had long since been sold, and what little cooking she did she managed on the two covers of the squat wood heater in the room upstairs.

Effie Anthony, a "birder" friend who lived in Bar Harbor, felt that Cordie should have a real cookstove where she could prepare proper meals for herself. Effie knew that one of her neighbors had such a stove stored in the barn. She acquired it at no cost, got another neighbor to load it on his pick-up truck, and they set out for Ellsworth.

But when they drove into the Stanwood dooryard they ran into a stone wall--the wall of Cordelia's will and determination as she stood in the doorway and barred their way. She hadn't asked for the stove, she didn't want the stove, and she would thank them to take the stove away and mind their own business. Hardly polite, and a strain upon friendship. But any act that even hinted at charity was utterly intolerable to her. Pride, yes, perhaps false pride, but she was a proud lady.

It is true that when nearly ninety she did have to accept state aid for the elderly, but through the efforts of Senator Eugene Hale, Governor Brann helped her uphold her dignity through generous checks for collections of her bird photographs for the state library.

June Forsythe, who lived across the road from the Stanwood home, did what she could to make life less complicated for Cordelia in her later years, while June's twin brothers, Herbert and John, saw to it that her woodbox was kept filled and that she had fresh water from the Boiling Spring each day. But, as June points out, Cordie herself would often precipitate crisis by impulsive or irrational acts. For example, on one particularly bitter cold February morning June had gone over to see how Cordie was getting along and found her washing her hair in a room so cold that her head was enveloped in a cloud of steam. Strangely enough she could do things like this and apparently suffer no ill effects. Endless hours in the out-of-doors in all kinds of weather had undoubtedly built up an immunity and indifference to such trivial discomfort. This, however, was small consolation for those concerned for her security and safety.

The stairway leading to the upstairs room where Cordie chose to continue to live in her independent way is narrow and treacherous. It projects half way across the doorway into the chamber itself and has a right-angle turn in the bottom steps leading to the dining room. Many marvelled that Cordelia had escaped serious injury on this dangerous stairway, but were even more disturbed by her careless use of kerosene lamps and candles. One night when going downstairs while holding a lamp in her hand, she had caught her heel on one of the treads and fallen. In the fall the lamp was broken, but luckily extinguished, and she escaped with only a sprained ankle.

And added to this fire hazard was another even more serious. Most of her firewood was spruce or pine, rich in resinous pitch which burned and sparked furiously. When Cordie chose to have a good fire she did not hesitate to fill the firebox to the top with pitch-laden slabs with the drafts wide open. Alarmingly soon the covers on the stove would turn a glowing cherry red as steam spurted from the iron teakettle.

One of the most bizarre stories about Cordelia's odd behavior during the years she lived alone is about what she did one cold day when she found herself short of firewood. She went out to the barn looking for pieces of board or whatever else would serve as fuel and came across an old tire from one of her brother Harry's early cars. Yes, indeed, rubber would burn! So, she carried it into the house and proceeded to cram it into the stove. Unfortunately it resisted her most strenuous efforts to push it all the way down into the firebox and she was unable to close the top. The fire department arrived in time to save the house, but thereafter Cordelia lived in a home with blackened walls and ceilings. She could not afford to do anything about it.

Snapshot by Ida Wilbur.

As she grew older, more of Cordie's eccentricities hardened into real problems. However much one might prefer to overlook or ignore it, there is evidence that Cordelia Stanwood suffered from something like a persecution complex in her later years. Her hermit-like seclusion and self-imposed isolation from others may have caused this. It is never good for one to be too much alone, and in this respect Cordie was unquestionably her own worst enemy.

One particularly strange delusion, and one which became almost an obsession with her as she grew older, was the conviction that meeting any person named Smith would automatically bring her bad luck. Ida Wilbur, a commercial photographer who assisted Miss Stanwood in her camera work and often travelled around the countryside with her on photographic ventures, relates that whenever they happened to meet some person on the road who was a Smith Cordie would promptly insist that they turn around and return home. They would not, she claimed, be able to get any good pictures on this day. And should a Smith happen to pass the house while she was out in the dooryard, she would quickly abandon whatever she was doing and shut herself inside for the rest of the day. In some long forgotten corner of her memory lingered a real or imagined unpleasantness caused by a Smith. It is hardly conceivable that the

One of the last photos of Cordie taken by Ida Wilbur.

non-speaking relationship between herself and her brother which went on for fifty years could possibly manifest itself so grimly because Harry had chosen to marry a Susie Smith.

Why did this woman who lived so freely and happily among the creatures of the wild become a victim of such superstitious nonsense about other people? How could such a strong and dedicated person in all her glorious independence become so vulnerable?

The answer must be sought in that loneliness engendered by her unrelenting refusal to walk the beaten track, a kind of loneliness which fostered eccentricity and distorted perspective. To know oneself and to accept the responsibility imposed by such self-knowledge is the major challenge of a lifetime. Cordelia, unfortunately, had achieved her self-command by ignoring the consequences in her social relationships.

"All the world is peculiar, except thee and me." One can say so, but one had better say so with tongue in cheek; for strength and weakness are relative for all. Those who might be inclined to resent this review of Cordelia's shortcomings and pecularities as inconsequential and unnecessary should pause to consider that every human being is a complex entity, a microcosm of good and bad, of strength and weakness, of greatness and mediocrity. Cordie was no less so than anyone else, but there is a consolation in the sure knowledge that her accomplishments would always outweigh her failings.

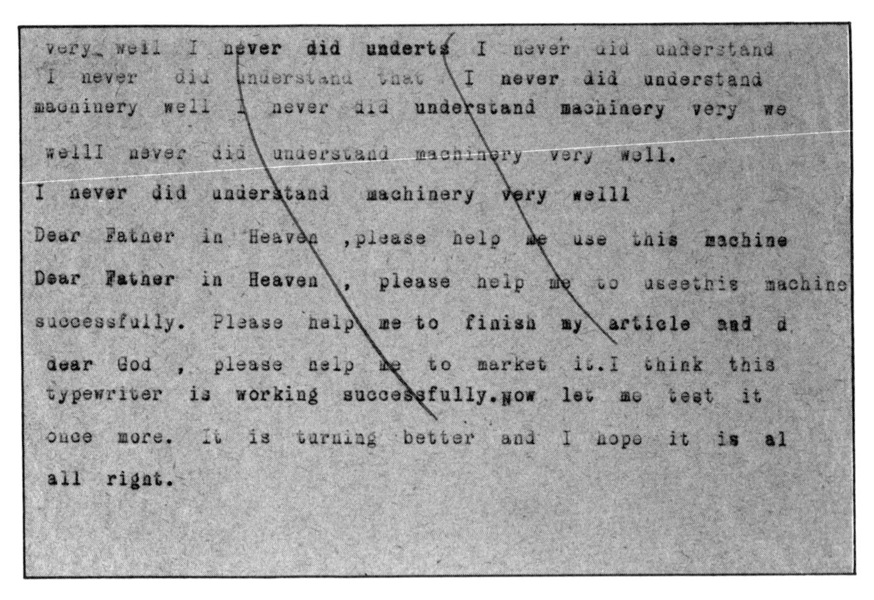
very well I never did underts I never did understand
I never did understand that I never did understand
machinery well I never did understand machinery very we
wellI never did understand machinery very well.
I never did understand machinery very welll
Dear Father in Heaven ,please help me use this machine
Dear Father in Heaven , please help me to usethis machine
successfully. Please help me to finish my article and d
dear God , please help me to market it.I think this
typewriter is working successfully.Now let me test it
once more. It is turning better and I hope it is al
all right.

This scrap of typing from Cordie's first efforts on her Hammond Multiplex machine speaks eloquently of her character.

13

Alone But Never Lonely

I have no life but this,
To lead it here.
Dickinson

Dramatic changes came to the Stanwood family on Beckwith Hill between 1910 and 1920. On Christmas day in 1911 Harry had married Susan Smith of Ellsworth Falls and was no longer a part of the household. Three years later in the summer of 1914, Captain Stanwood died, preceding by

The Stanwood Homestead in about 1915, with the new dirt road up Beckwith Hill. (CJS photo)

The dooryard of the Stanwood homestead, about 1915. (CJS photo)

only one year the passing of the Providence aunts. Cordie, in her own private way, felt these losses keenly but found her greatest comfort through even more intensive effort in her woodland studies.

Then, in 1915, Aunt Addie, the Captain's oldest sister, ninety-four and a widow, came back from the west to live with Cordie and her mother. She brought with her the same metal-studded trunk which had gone with her over the Oregon Trail in a covered wagon fifty years before, and when the train pulled in at the Ellsworth railway station she had danced down the length of the station platform with a much startled conductor. Aunt Addie lived to be ninety-nine.

The nineteenth century was quickly fading into history, and her sister Maria's son Alfred was the last of that Stanwood line settled by Job on Cranberry Isle back in 1760. Cordelia accepted this inevitability of change, but after her mother died in 1932 there was no one left to share in her life other than her sister Della and a few friends.

It cannot be denied that the last eighteen years were difficult for Cordie. She had exhausted her financial resources despite her frugal habits, and the earning capacity of a woman over seventy-five was meager indeed for one reduced to walking countless miles throughout the countryside to sell stationery and Christmas cards. Help from her brother Harry was, of course, out of the question, and her sister Della had lost her husband and was supporting herself by working in the art department of a large store in Worcester, Massachusetts. What little help Della could afford had to be handled very circumspectly, as the milk delivery and the cookstove fiascos made plain. Cordie was not the kind of person for whom one could do very much to make life more comfortable and secure. Her proud spirit abhorred any charitable act as a threat to her independence.

In those last years by herself she might have become very lonely had she not been so completely involved in the eternal rhythms of life all about her in the ever-changing sea sons. The world of nature had been her salvation. That friend she had "associated with since childhood days," that friend she had "summered and wintered with" had not failed her.

Although Cordelia Stanwood lived on for a quarter century beyond the normal three score and ten, neither her solitude nor her isolation from the rapidly changing world around her could ever touch that inner joy and peace of mind she had found abundantly among her birds. Every hour brought its own special promise, and she met the challenge of each new day with the same inexhaustible enthusiasm that had led her surely to her greatest achievements.

Both Emerson and nature had taught her well, and there was no loneliness in being alone. The continuity of life surrounded her in the fragrance of flowers, the song of birds, the glory of a sunset, and every dawn brought fresh adventure.

In the mud and scum of things
There alway, alway something sings.

Epilogue

The old homestead on the hill was wrapped in quiet stillness when Cordie was no longer there. Grass grew tall in the dooryard, and the Japonica, the lilacs, the rugosa rosebushes, the honeysuckle, and the mock orange reached out across the doorways and the windows to shut it away from curious, staring eyes. This was a home filled with long long memories, memories which sent whispering echoes through empty rooms from cellar to attic.

Proudly neat and trim when Cordie was born, it had grown old as she grew old and now stood alone, silently dignified but forlorn as if waiting for a step or voice to bring it back to life. The marks of neglect were upon it. It needed paint; door hinges sagged, and windows rattled in their casings. Squirrels had chewed their way inside under the eaves, and sharp teeth had shredded patchwork quilts for nests in bureau drawers. The rumpled corpse of a ruffed grouse lay under the dining room table where it had fallen when the bird crashed through a windowpane.

I had been made sadly aware of what had happened to Cordelia's home when I first visited it on Sept. 21, 1958, with Helen Raymond, who had been appointed her guardian after she had become, at the age of 89, a ward of the state. One part of me ignored or rejected the neglect and deterioration my mind was unwilling to accept, but another part of me was excited and stimulated by the rush of memories stirred up by finding myself so close to the physical center of Cordie's life. In fact, for two memorable hours the Stanwoods were home again. In my mind's eye I saw Grandmother Stanwood sitting in the Lincoln rocker beside the old square piano in the parlor, heard Maria teasing Della on the front hall stairs, smelled the newly baked bread Mother Stanwood had just removed from the Dutch oven in the dining room, watched Cordie feeding a young thrush by a window with a piece of steak she had caught up on the point of her scissors. The faded rugs, the nineteenth century furniture, the framed pictures and old lamps, the marble-topped bureaus and patchwork quilts had brought it all back most vividly.

What was to become of it? How save the stuff that "dreams are made on?" Could it be salvaged from the fate which had already overtaken so many of the old landmarks that had vanished in the name of progress? I knew I was not alone in my concern over the fate of the Stanwood home, and others, like myself, were fully aware of the problems confronting anyone who might wish to acquire it.

Land values were climbing, and the west slope of Beckwith Hill was a choice site coveted by promoters for a motel complex. Although this placed the property in a highly competitive market an even more serious obstacle was raised by the heirship rights of surviving members in the family who could not agree among themselves about what to do. Furthermore, since Cordie had received old age assistance, the State of Maine had a claim against the estate. After Captain Stanwood's death in 1914 the house itself had been sadly neglected and the large barn was beyond repair--the roof had fallen in.

Only one and four-tenths miles lie between the traffic light at Main Street in Ellsworth and the driveway to the Stanwood home on Beckwith Hill, and since Miss Stanwood's death in 1958 the dwellings, the open fields, and the partially wooded pastures had all but vanished before a steady sprawl of commercial development. Concrete, steel, and asphalt had replaced everything familiar. Cordie herself, even before she went to the nursing home in 1955, had become agitated by these impending changes which threatened to alter or destroy everything which she had loved so dearly. She missed her fields and pastures and resented the increasing clatter and rumble of traffic up and down the hill.

This was the general situation on January 6, 1959, less than two months after Cordelia had been buried in McKenzie Cemetery beside her parents, when I was asked to speak to the Ellsworth Rotary Club. Knowing that several of the older men in this organization had known Miss Stanwood all their lives and had some knowledge at least of who she was and what she did, I felt this to be a propitious opportunity to tell her story and give her that recognition she well deserved but never enjoyed while she lived. Also on my mind at this time was the fact that an ad had just been placed in *Yankee* magazine offering the Stanwood place for sale.

It would be dishonest of me to deny that I was a very strong advocate for Cordelia Stanwood in what I had to say that night, and the results afterwards were quite unexpected and highly gratifying.

Hervey Phillips, who had been successful in the business world and was then president of a local bank, listened attentively to my review of Cordie's accomplishments and when I had concluded he spoke up frankly to urge that something be done to save the Stanwood property. When I made it quite clear that this had been my own dream for the past year, Hervey startled me with a comment which would irreversibly change the rest of my life: "Go after that place, Chandler. I will back you personally to do it." Although prior to this I had been able to arouse considerable interest in Ellsworth's bird woman, this was the first time anyone had offered that essential financial support so necessary to make it come true.

I cannot find the words to express my gratitude to Hervey Phillips. My vision without his backing would have been futile, and except for him the Stanwood Wildlife Foundation would still be only a dream and the homestead museum merely a memory.

The availability of purchasing power to acquire the Stanwood homestead, however, did not in itself make this an inevitable conclusion. There were other matters to be resolved, involving money but complicated by knotty legal problems and unanswered differences of opinion among surviving members of the family.

First of all it was essential that the necessary steps be taken to liquidate the state's claim against the estate for the modest pension, medical assistance, and nursing home care which Cordie had received over a period of five years. Once this was resolved I could approach the heirs with a concrete proposal. Della, the one remaining sister, now a widow in her eighties but still working in Worcester, had already been very helpful to me in my biographical research and was favorably inclined toward my earnest desire to acquire the Stanwood property.

Brother Harry, on the other hand, and Maria's son, Alfred Langewald, were less predictable. Alfred, who lived in Pennsylvania, had been out of touch with the family for several years, and Harry was bedridden with the same malady that had taken Cordie's life--cancer.

Knowing so well how much bitterness and ill-feeling had kept Cordie and Harry apart for so long, I hesitated to approach him, but fate stepped in to solve the problem for me.

In the late spring of 1959 the homestead was entered and Cordie's typewriter removed from the premises. A neighbor who was aware of my interest in Miss Stanwood called me to report the incident, and I promptly notified the police, adding a suggestion as to where the machine might be found. The information proved correct, and within two hours the typewriter was back where it belonged.

What otherwise might have been the happy conclusion to a disturbing incident turned into a real crisis when someone's phone call to Harry's spirited wife Sue informed her that a certain person named Richmond had been to the Stanwood place doing something or other with the police department.

Sue, who could speak her mind with no prompting, demanded to know what business I had there and what it was all about. Quite clearly she and Harry were entitled to an explanation; so, with considerable misgiving and a large measure of uncertainty I drove out to Tunk Lake in Sullivan to see them.

It is amazing how much feeling can be put into one word. Sue met me at the door with one such word:

"Well?"

There was but one way, I told myself, to meet this situation. Be completely frank and honest and hope for the best.

When I had explained why I was interested in Cordie,

what I hoped to accomplish with her life story, and how much it would mean to establish the homestead and the land around it as a living memorial to the Stanwood family, all suspicion of me seemed to evaporate, and both Harry and Sue became friendly and responsive. In fact, I was invited to join them for one of Sue's famous Down East meals which had attracted sportsmen to Big Chief Camps for forty years.

After dinner during Harry's nap-time I took a walk around the Big Chief Campgrounds, and when I returned Sue told me that they had talked it over and come to a decision. They would go along with me in my efforts to gain title to the Stanwood property and would even write to Alfred in Pennsylvania urging him to do likewise. Harry was failing rapidly, and I believe he honestly felt it was time to forget old enmities and do something which would do credit to the Stanwood name.

So much had happened and so suddenly it was staggering. I felt that I should sit down in some quiet place and think about it and give myself time to assimilate all the implications, but there was no such time available. Everything was moving, moving rapidly and surely toward the acquisition of that which had seemed completely unattainable only six months before.

By the end of the summer all deeds were signed and recorded, and in November it became my happy privilege to transfer Birdsacre's forty-odd acres and the one-hundred-year-old homestead to the newly incorporated Stanwood Wildlife Foundation.

Dr. Clarence C. Little, former president of the University of Maine and founder of the Jackson Laboratory in Bar Harbor, was our first president, while Helen Raymond's son John served as vice-president, Charles J. Hurley of the law firm of Hale and Hamlin acted as clerk and legal counsel for the corporation, and I was named curator and manager. The other members of the original board of trustees were June Forsythe; Reverend Margaret Henrichsen, author of *Seven Steeples*; Agnes Huston, founder of the Cordelia Stanwood Bird Club; Grace Radloff, and Dr. James Crowe.

This small group, greatly assisted by other groups and many enthusiastic volunteers achieved unbelievable results between November and the following summer. The skeleton of the old barn was razed and burned by the Ellsworth Fire Department to make way for a parking lot and driveways put in by the Sargent Construction Company. The gray, weathered clapboards of the house greedily absorbed some fifteen gallons of high grade paint furnished and applied by Rotarians, and for the first time in fifty years the homestead began to resemble the proud structure it had been when Captain Stanwood came home from his sea voyages. Inside the house the work of renovation and restoration went on more slowly but steadily all through the winter, and many willing hands were there to help. Chief among these was Mac Sawyer, an accomplished carpenter who became my closest associate in deciding what had to be done and determining how to do it. Peter Hoit, a painter and former neighbor of Cordie's, came forward to do most of the woodwork and all the floors in the house, while Agnes Huston's husband Frank joined me on the coldest day of that winter to paper all the downstairs rooms.

These were not young men, but they were masters of their trades, and they, together with many others less skilled but equally willing, made it possible for us to open the Stanwood Homestead Museum officially to the public on Cordelia's birthday, August 1, 1960.

It was a memorable day, a day of celebration, a day of dedication and memories, and, above all, a day of high hopes for the future. Here Cordelia Stanwood had reached out "beyond the spring" to achieve her life's dream, and her beloved Birdsacre and her home would now be here for all to share in every spring beyond this day.

All through that day after the doors to the homestead were thrown open to the morning sunlight until the last visitor had signed the guest book, the spirit of the Lady of Birdsacre seemed to be everywhere. For almost a century she had been so much a part of all that had ever happened in this house or its immediate surroundings that one felt her presence at every turn.

Inasmuch as I was more involved than others in the background of events that had led to this memorable day, I felt her nearness in a special way. How right that Cordelia's home and her woodland sanctuary should reflect her vital personality so intimately, and all that has ever been done over the years since this dedication day in 1960 has only served to emphasize and reaffirm the character of the remarkable woman who had lived here.

The Stanwood Homestead Museum today, open to visitors.

Chandler Richmond, curator and manager, at the entrance to Birdsacre, with Ollie, the barred owl who is one of the current tenants of Birdsacre.

The years have slipped away all too swiftly for me since 1960 when I became so happily involved in the development and expansion of the Stanwood Homestead Museum and Birdsacre Sanctuary. There is an intimacy and vitality about the place that radiates the living spirit of the woman who lived here. Whenever I walk upon the trails worn deep into the forest floor by the footsteps of Cordelia Stanwood, I am reminded that all of us, in one way or another, are engaged in the same search; and to learn how well she ultimately succeeded is a source of encouragement to everyone who seeks to broaden the horizons of our understanding.

Birdsacre is a natural resource where all may find those truths which had enriched the life of Cordelia Stanwood. It remains essentially as it has been for over a century. Trees grow, reach maturity, and sink back to earth to build the humus for new growth. The flowers, ferns, and shrubs that thrust their way skyward among the forest trees are no different from those found elsewhere, except that they represent the same perennial abundance that struck the eye and inspired the pen of the Lady of Birdsacre. Our visitors walk the same trails she walked, see and hear the same birds she found, and are welcome to enjoy that same pleasure she knew when away from the paved highway.

Each spring when the snows melt, Birdsacre becomes vibrant with plans and projects for the coming season. Some are part of that essential restoration work carried on continuously, but many involve extension and expansion of those features which have made the homestead and sanctuary so appealing to our many visitors.

The record is impressive and becomes more so with each passing year:

1960 Old trails cleared and connecting trails opened. A central heating system installed in the Homestead.

1961 Parking lot inclosed by a fence built with 100-year-old cedar rails. Registration lodge constructed.

1963 New timbers under the front of the main house.

1965 Celebration of Miss Stanwood's one hundredth birthday.

1966 Bown Pond created and stocked with trout.

1967 Purchase of additional acreage through Ruth Foster gift. Highway by-pass crisis averted.

1968 Homestead re-shingled by the Ellsworth Rotary Club.

1969 McGinley Pond created.

1970 After one hundred years the Homestead has running water and a bathroom.

1971 Combination windows provide additional insulation. A new heating system installed.

1972 Latticed green blinds added. Annie Reich Memorial Pool created through gift by Eva Reich Moise.

1973 Homestead placed on National Register of Historic Places. Outdoor bird shelters constructed.

1974 Discovery of Stanwood photographic plates by William Townsend and return to Birdsacre by Acadia National Park. Property lines surveyed.

1975 Creation of Martinland — memorial to Alfred G. Martin. Presentation of Martin oil paintings by Erastus Corning II.

1976 Winter House built.

1977 Security system installed.

1978 The wildlife rehabilitation center releases about 50% of the birds brought in each year. Those incapable of independent survival remain at the Sanctuary to educate and delight visitors and to serve as models for photographers and artists. Especially popular is Ollie, the barred owl, who lived until 1989, and his 20th year.

Sarah Doyle, Cordie's English teacher, consistently admonished her pupils at the Girls High School in Providence to remember that "the eye sees what the mind brings the power to see." I have observed that power reflected in the mind's eye of Cordelia Stanwood through countless pages of her writings, and to me she is still a living force in and about her home as she was in life. From room to room she is beside me when decisions are to be made about arrangements of furniture, the hanging of a picture, even the location of a lamp. Her vital spirit and the force of her personality are everywhere, and always most clearly when I walk upon her trails in the woodland. We share that slant of sunlight across the path which seeks out the first blossom of the arbutus and listen together to the thrush's song in the shadows beyond the spring. I am never alone when I watch the busyness of honeybees about the hollow trunk of the old hemlock, for these are the offspring of Cordie's bees from those hives behind the barn when I was still a boy, and the sharp "Wicka, Wicka, Wicka" of the flicker at the domed ant hill by the edge of the orchard is a vocal testimonial to the continuity of her world. This is Birdsacre; this is Cordelia's home. She is still here. I like to feel that it will ever be so, and that she will always be here for others as she is now for me so that all might see what she saw so vividly--that both mankind and nature are forever linked in the great and wonderful mystery of creation.

Postscript

"an addition appended to a completed book."
Webster

In large measure, Beyond the Spring has neither beginning nor end. It is an ongoing chronicle of achievement impervious to time. Cordelia set it in motion, others carry it forward, family heritage, fortuitous location, and selfless commitment have made Birdsacre what it is today, and only in the future shall one find its ultimate destiny for good as envisioned in Cordie's passionate quest.

A quarter century ago, when I first began to seek out those cherished areas in the woodland so often referred to by Cordelia in the field notebooks, it was a constant adventure of discovery. Then did I find the fragrant Twin flower broadcasting its delicate perfume for me, as it had done for her three score years before.

In her footsteps I came upon the thick mats of trailing arbutus on the Dorgan Hills a half mile to the west of the Homestead, and on the south boundary of the Weinstein Twenty -- long before it became a part of Birdsacre -- I found the tall pines of the Queen's Throne where she had so often listened to thrushes at eventide.

All now integrated within the Sanctuary, now the "forty odd acres more or less" have become one hundred. It is our protective shield against change and a rich legacy for the future.

There are many ways whereby one might hope to capture the elusive fascination and complexity of Cordelia Stanwood. Her idealism, her endless patience, her unshakeable determination to see, and hear, and know all that was there to find were constantly reflected in her writing, and then one day I came upon a scrap of paper among her manuscripts which says it all most graphically.

She had bought herself a Hammond Multiplex typewriter, and much as she had done previously with the cumbersome box camera Frank Chapman had secured for her, she set out to master its mechanical mystery to meet her mood.

The mind and spirit of the Lady of Birdsacre is right there in every word tapped out so painstakingly on the keyboard of her new typewriter on that day so long ago.

> very_well I never did underst I never did understand
> I never did understand that I never did understand
> Machinery well I never did understand machinery very we
> Welll never did understand machinery very welll
> Dear Father in Heaven please help me to use this machine
> Dear Father in Heaven, please help me to useethis machine
> Successfully. Please help me to finish my article and d
> dear God, please help me to market it.I think this
> typewriter is working successfully.Now let me test it
> once more. It is turning better now and I hope it is al
> all right.

Fifteen years have slipped away since the Stanwood Homestead was placed on the National Register of Historic Places, and the present preservation of the Homestead to its original condition has been generously supported by a matching grant through the Historic Building Bond Issue approved by the people of Maine on November 5, 1985.

Cordie's birthday, August 1, is now officially recognized as Cordelia Stanwood Day in Ellsworth.

Legislative action by the State of Maine also recognizes Birdsacre as a state protected refuge formally known as the Stanwood Wildlife Sanctuary.

My goal, shared by all who support us, was recognition for Cordie.

She has it.

Chandler S.W. Richmond, 1988

P.S.S. 2007

These footnotes to the future verify that hope expressed in the Epilogue of 1978 and the Postscript of 1988, that both time and circumstance would be kind to Cordelia, her home, and her beloved Birdsacre. Time and circumstance seem only to improve the inherent value and uniqueness of Cordie's Sanctuary, as the world beyond her borders ever alters so drastically. The character of Birdsacre in the 21st century is rare and precious, dedicated to remaining intimate, simple, and sincere, where one may find sanctuary in the gentle, timeless rhythms of the out of doors. That quality of character that imbued the soul of Miss Stanwood lingers on in her legacy and radiates within all those who have divined an equal sense of its rarity here.

Today, Birdsacre has grown into more than merely a memorial to a great lady's life and achievements. The vitality of her world and all she loved continues to grow and thrive; and, surprisingly, the visiting public discovers an endless variety of unanticipated ways of enjoying this special space. Youth groups, outreach programs, and children explore, play, and learn; local clubs and groups from Master Gardener Volunteers to the Downeast Audubon congregate, renew, and contribute; while artists of all varieties seek subjects and inspiration, teaching occasional courses on site. From birthday parties to weddings, from cross-country running to snowshoeing, Birdsacre's endless enjoyment ever expands. Near the character core are the educational nature programs, rehabilatory assistance, and the Sanctuary's ability to accept, protect, and monitor land grants and conservation easements, which one senses would all earn her approving nod.

Cordie still catches kudos, too. From the reemergence of her long lost photographic glass plate collection, discovered by Bill Townsend at the Acadia National Park in 1974, to Authur Laurent's recent French publication at the Bourges Museum d'Histoire Naturelle, Cordelia's achievements seem eternal, and her story shared without end. The 1974 discovery may prove to be the sole surviving complete collection of turn-of-the-century ornithological photographic glass plates to date. Long ago Cordelia sold her original box camera, but through the aid of Kodak, a duplicate Eastman No. 5 resides in the house once more.

In 1978, Chandler Richmond's biography *Beyond the Spring* was the first publication to tell Birdsacre's story. In 1981, Ollie, Birdsacre's popular resident Barred Owl was the subject of a fictionalized story by Ada and Frank Graham, in which Ollie and Chandler were featured in *Jacob and Owl* by photographers Frank and Dorothea Stoke. In 1982, Ada Graham followed with *Six Little Chickadees: A Scientist and her Work with Birds*, a children's book of Cordelia's studies.

More recently Cordelia has appeared in the chapters of these authorized publications: *Bird watching with American Women*, by Deborah Strom, 1986; *Women in the Field*, 1991, and *American Women Afield*, 1995 by Marcia Meyers Bonta; and *Pionniers de la Photographic Animaliere* by Arthur Laurent in 2006, confirming Cordelia's pioneer position as the first professional female ornithological photographer.

In 1982, Stan Richmond joined the Birdsacre family, and in 1990 assumed the role of caretaker. For twenty-five years he has cared for the non-releasable residents, maintained the woodlands, presented educational owl programs to the community, and entertained the public with his humorous stories.

Raising of the Richmond Nature Center in 1992 provides a public gathering/display area for an excellent nature collection consisting of bird mounts, nests and eggs. This multifaceted structure also houses the Clover Morrision Library, showcases local artwork, and provides an unusual, variety gift shop. Through the Beautify America Grant, a

3 tier, handicap assessable porch patio grew behind the Nature Center, and was designed by the inventive, Ellsworth 3rd and 4th graders for the educational owl programs schools receive there.

More recently, Diane Richmond Castle has breathed life back into the inner sanctum of Cordie's Homestead. Her personalized tours, intimate touches, cut flowers taken from blooms about the house, and, with her husband Arthur, printed individual editions of Cordie's various bird life chapters have both kept Cordie's memory vibrant and fragrant.

The remaining 40-acre Stanwood property has grown to protect over 200 acres of the woodland Cordie traversed. Today the perimeter trail encompasses the whole, leaving much of the interior wild.

Cordie' s world is alive and well, nurtured and carried on with great care.

Grayson G. D. Richmond, grandson

The Aegis of Birdsacre, accepts, protects, and monitors land grants. Over the years, conscientious individuals have supported Birdsacre's land conservation attempts with memorial contributions preserving the outdoors for our enjoyment and that of future generations. Long ago, off today's Red Trail, Cordie's brother Harry Stanwood carved his initials in a rock for the future to remember him. Following his example various ponds, pools, favorite spots, land acquisitions, gardens, trails, aviaries, and benches have been memorialized bequests honoring loved ones.

1959 The Homestead	Property acquired through backing of Hervey Phillips.
1960 S.W.S.	Stanwood Wildlife Sanctuary Incorporated.
1962 Bown Pond	Named for Cordelia's maternal family, honors the Stanwood family's gift in memorializing Cordelia.
1968 Foster Trail and Land	Honors Pat and Representative Ruth Foster's support.
1970 McGinley Pond	Dedicated to George McGinley, a State biologist involved with testing DDT in Eagle eggs.
1972 Annie Riech Pool	Honors the memory of Eva Riech Moise's mother.
1976 Martinland	Dedicated to the memory of loyal supporter, naturalist, artist, taxidermist, and writer Al Martin.
1980 Hazel Walker Aviary	Recognizes close friend of Cordelia Stanwood and active supporter of Birdsacre.
1981 Richmond Pond	Surprise dedication, recognizing Marion and Chandler Richmond as founders of Birdsacre.

1982	Weinstein XX	Celebrates Frederic and Suzanne Weinstein's aid in saving the 20 acres surrounding the Queen's Throne.
1983	Mitchell Lot	Acquired by 13 supporters.
1984	Harriet's Pond	In honor of Mrs. Bayard Ewing.
	Dorgan Lot	In memory of Mary McFarland.
1985	Children's Acre	Remembers the devoted efforts of Ruth Hanna.
	Jones Lot	Donated by a childhood friend of Cordie, Beatrice Jones.
1989	Jordan Lot	Neighbor, Robert C. Jordan's lot honors Hazel Walker.
	Whitcomb Woods	Named for Paul Whitcomb's generous bequest.
1992	Herpel Path & Pond	Honors Professor and Mrs. George Herpel.
	Hazel Stanwood Chase West	Bequest and donation of family Heirlooms.
1995	Union River Dam Lot	Donated by Bangor Hydro-Electric Company.
1997	Western half of Jordan Lot	Carroll and Alice Jordan's contribution with brother Robert's makes up 40% of the Sanctuary.
1998	Woodland Gardens and Boardwalk	Master Gardeners Program, directed by Bill Booth, and in memory of trustee, Russell Kittridge.
2000	Pinkham Path	Remembers Albion and Betty Pinkham, long standing friends.
	Pinkham Gazebo	In memory of Jean Pinkham.
2001	St. Francis Garden	Celebrates the Kittredge family's long commitment.
2003	Endowment Growth	Dorothy Gaspar's wise financial guidance and friendship.

2004	Stevenson Bog	Recognizes neighbors Andy, Becky, Lyle, and Nate Stephenson.
2005	Sullivan Lot	Jim Allen donates a shore lot on Taunton Bay.
2007	Stony Brook Road Lot	In memory of trustee, Eugene King and wife Dorothy E. King.
	Engraved Granite Bench	Mary and Henry Wider, given by Susan and Sarah Wider.
	Engraved Granite Bench	Remembers Grace, Winnie, and Marion Lord.
	June Skinner Wood Burned Bench	for extraordinary children's guide, Mona Shea.
	P. Weirs & S. Jordan Painted Bench	for devoted gift shop volunteer, Charlotte Kopfmann.

Clover Morrison Library

Dedicated to Clover Morrison	Member of Cordelia J. Stanwood Bird Club, Homestead guide, and generous Sanctuary benefactor.
Ned and Peg Latham	Bequest and donation of their library to Birdsacre.
Inez Boyd	Bequest and donation of extensive nature and bird collections. Energetic supporter and Stanwood Wildlife Sanctuary President.

Index